Hazardous Waste Management

Volume I
The Law of Toxics and Toxic Substances

Editors

George S. Dominguez
Director, Information Services
Synthetic Organic Chemical
Manufacturers Association
Washington, D.C.

Kenneth G. Bartlett, Esq.
Associate Editor
Toxic Substances Journal
New Haven, Connecticut

CRC Press, Inc.
Boca Raton, Florida

Library of Congress Cataloging-in-Publication Data
Main entry under title:

Hazardous waste management.
   Bibliography: p.
   Includes index.
   Contents: v. 1. The law of toxics and toxic
substances.
   1. Hazardous wastes--Law and legislation--United
States. I. Dominguez, George S. II. Bartlett,
Kenneth G.
KF3946.H39   1986   344.73′04622   85-29889
ISBNM 0-8493-6356-X

Direct all inquiries to CRC Press, Inc., 2000 Corporate Blvd., N.W., Boca Raton, Florida, 33431.

© 1986 by CRC Press, Inc.

International Standard Book Number 0-8493-6356-X

Library of Congress Card Number 85-29889
Printed in the United States

THE EDITORS

George S. Dominguez, Director of Information Services for the Synthetic Organic Chemical Manufacturers Association (SOCMA), received his B.S. and M.B.A. degrees from Kentucky Christian University. Mr. Dominguez held various Product Management positions with Ciba-Geigy Corporation from 1953 to 1971 and became Director of Government Relations, Safety, Health, and Ecology in 1975. He then moved on to Springborn Regulatory Services where he served as President from 1981 to 1984.

Mr. Dominguez has written numerous journal articles, books, and pamphlets and is currently Senior Editor of the *Toxic Substances Journal.* He has also presented many technical lectures on the subject of toxic substances. Mr. Dominguez is a member of the editorial boards of the *Regulatory Analyst* and the *Environmental Newsletter,* and is also a member of the International Professional Association for Environmental Affairs.

Kenneth G. Bartlett is an experienced Trial Attorney. He has been certified as a civil trial specialist by the National Board of Trial Advocacy. He maintains a practice in New Haven, Connecticut primarily focused in environmental law and general personal injury litigation.

He received his undergraduate degree in Geology from the University of Vermont in 1977 and while at Vermont was the recipient of a National Science Foundation (NSF) Undergraduate Fellowship in Trace Metal Analysis. He received his Juris Doctor from the University of Bridgeport School of Law in 1981.

He has been actively involved in numerous environmental legal issues, including formaldehyde and hazardous waste litigation and a variety of "Toxic Tort" cases. He is admitted to the Connecticut and New Jersey Bars and is a Contributing Editor of the *Toxic Substances Journal.*

He has published numerous articles on environmental law and has lectured widely, including the Association of Trial Lawyers. He has been a member of the faculty for ATLA's Advanced College of Advocacy Course on "Toxic Torts".

CONTRIBUTORS

Stephen Bailey, M.S.
Professional Engineer
ICF Incorporated
Washington, D.C.

Kenneth G. Bartlett, Esq.
Associate Editor
Toxic Substances Journal
New Haven, Connecticut

David A. Bruce, B.A.
Attorney
Maryland Bar
American Bar Association
Washington, D.C.

George S. Dominguez
Director, Information Services
Synthetic Organic Chemical
  Manufacturers Association
Washington, D.C.

Richard deC. Hinds
Partner
Cleary, Gottlieb, Steen & Hamilton
Washington, D.C.

Jeffrey H. Howard, Esq.
Attorney
Davis, Graham & Stubbs
Washington, D.C.

Edward W. Kleppinger, Ph.D.
Principal
EWK Consultants Incorporated
Washington, D.C.

Gilah Langner, B.A.
ICF Incorporated
Washington, D.C.

Mary Jean Marvin, M.A.
Staff Attorney
Environmental Law Institute
Washington, D.C.

Robin Rodensky, M.S.
Senior Associate
ICF Incorporated
Washington, D.C.

TABLE OF CONTENTS

Chapter 1

HAZARDOUS WASTE MANAGEMENT

George S. Dominguez

TABLE OF CONTENTS

I. INTRODUCTION

Literature on hazardous waste has grown enormously in recent years. There seems to have been an almost endless succession of books, articles, and reference materials, not to mention conferences, seminars, and a virtually endless stream of government publications dealing with various aspects of this timely subject. These, prodigious as they are themselves, do not include the media coverage of virtually every aspect of hazardous waste from considerations of public health and safety to elaborate and exhaustive editorials soundly and roundly criticizing industry for its past and present actions on the one hand and government for its lack of action on the other. Since the essential purpose of this guide book is to provide working managers with a comprehensive introduction to practical operational aspects of hazardous waste management and with an extremely important foundation in revelant laws, rules, and regulations, it is not our place nor intention to enter into the debate of the relevant responsibility or lack thereof of past or present actions of the private or public sectors. Rather our purpose is to concentrate on those aspects of hazardous waste management which we feel either have not been previously addressed, insufficiently examined, or most importantly not considered in an integrated fashion.

II. THE HAZARDOUS WASTE PROBLEM

Since so much has already been written about the nature and extent of the "hazardous waste problem", there is little purpose served in attempting to restate the entirety nor the complexity of the problem; to a great extent it is, in fact, a self-evident problem. There are a few points, however, that do warrant mention since they help to place present actions and future activities and the potential for adequate waste management control techniques to be utilized in perspective:

1. The hazardous waste management problem is not new and in fact waste disposal problems have plagued mankind since introduction of even the most rudimentary attempts at manufacturing; numerous historical examples exist from the predawn of civilization as we know it to present times.
2. There is an obvious national as well as international dimension to the problem even though, unlike more conventional pollution problems such as air or water pollution, hazardous waste problems tend to be more localized since we are not dealing with the opportunity for wide-spread transboundary distribution phenomena that can occur in the air or water pollution situations. However, we are potentially dealing with ground and aquifer transfer situations and, therefore, while the extent of the transportation problem may not be great, it nevertheless cannot be totally discounted. In fact, one of our most serious problems can be ground water contamination which is not only difficult to measure adequately but difficult to predict, control, and obviously remedy once it has occurred. The other potential transfer situation of importance occurs when hazardous wastes are deliberately transferred from point of origin to point of disposal, hence a concern with spills during transport and with the public policy issues associated with the establishment of dump sites. Similarly, this transportation and disposal siting problem is international in scope since there have been numerous instances of countries generating hazardous waste and then transporting them to other nations for eventual disposal.
3. The problems with inadequately treated or inappropriate disposal of hazardous waste can, as we have all seen, take years if not generations to manifest themselves. Once manifested, remedial action is extremely expensive, time-consuming,

and in some cases arguably impossible. The practical significance of this has to be emphasized as various hazardous waste management options such as reduction in waste generation, waste treatment, and finally appropriate disposal techniques are considered.

4.   Identification and classification is important since it is the basis upon which various restrictions might apply. In addition, it is, of course, equally if not more important because it provides an indication of the potential application of the various manufacturing techniques already mentioned.

5.   Cost is undoubtedly a major consideration, either in system review, design, or management, to say nothing of actual capital and operating costs involved in actual treatment and disposal. However, as we have repeatedly seen in the media the costs of reclaiming a hazardous waste dump site, and the liability that now exists based on such status as RCRA and Superfund also impose external cost considerations that have to be taken into account when undertaking an analysis of the costs vs. benefits of various management and control options. Cost today, therefore, just like the various scientific and technical considerations as well as management options to be reviewed, has to be considered in its totality. We have to consider all aspects of cost — both institutional and extrinsic when considering the development of an effective and comprehensive waste management system. Also included in this calculus, as difficult as it may be, is that very important factor, namely, public attitude and response to improper waste management practices. These are reactions which, as we have seen, not only influence such decisions as the legislative adoption of laws like RCRA and Superfund but the actions of federal and state agencies in their implementation and enforcement actions. In addition, such public reactions can well effect the position of individual companies in the market place with consequences such as adverse effects on marketing of goods and services and even effects on stock values. All together, a new dimension of concern arises as technical managers must also take into account factors which historically have been considered outside their traditional realms of activity. The logical outgrowth is the development of a team approach to hazardous waste management — a concept which we will be describing in greater detail later in this chapter as one management option available to assure full integration of all of those diverse organization needs in hazardous waste management decisions.

As important as all of these points are, perhaps the most important of all is the necessity to recognize two major realities that have a direct bearing on the entirety of the hazardous waste problem and of our opportunity for future remedies:

1.   As has often been said, we are living in a highly industrialized technology-dependent society and on both a national and global scale, we are continuing to evolve into, if anything, an even more industrial and technology-dependent society. The benefits are obvious as we are surrounded daily by an unprecedented plethora of goods and services that are directly and indirectly the result of this industrialization and our technology-dependent society. One price that we must pay is waste generation. This is the first of the two major points that we must recognize if sound hazardous waste management programs are to be developed and applied. The principle is a simple one; namely, that as goods are produced, wastes are generated. While a great deal has already been written about so-called "low- and non-waste technology" and a great deal more will undoubtedly be written about this approach, it remains, at least for the moment, largely an unrealized goal and one which, while it holds great promise, will never totally solve

Table 1
PRIMARY SOURCES
OF HAZARDOUS
WASTES

Industrial
    Chemical manufacturing
    Manufactured products
    Formulating
    Extractive industries
        Mining
        Oil and gas production
Municipal
    Waste treatment facilities
    Power generation
Hospital
    Infectious and biohazardous
Reclamation
    Land and buildings
Nuclear
Agricultural

the problem because it is unrealistic to believe that the application of such techniques can be universally applied to all manufacturing and processing situations. The first point, then, is that we cannot sustain production of industrial goods and the numerous manifest benefits that derive from them without waste generation.

2.   There is a basic misconception in the minds of many that hazardous waste generation is, if not solely, predominately the result of industrial manufacturing and processing operations. In fact, this is absolutely incorrect. Hazardous wastes are generated from a number of public and private sector sources, some quite obvious and some, despite the potential hazards they generate, often totally unrecognized. To put this important consideration into perspective is essential since it is not only illogical but extremely ineffectual to consider from a public policy or a technical viewpoint control measures that apply solely or predominately to the industrial sector while overlooking or only minimally addressing these others. Table 1 lists the other important generators of hazardous wastes from sources which must be considered in any complete assessment of hazardous waste generation and management. It is only by examining these other sources that the full magnitude of the hazardous waste generation and management problem can be understood and adequate management techniques developed and applied. In this series we will be examining each of these sources in greater detail.

III. MAJOR AREAS OF CONSIDERATION IN HAZARDOUS WASTE MANAGEMENT

From the preceding brief introduction to the scope and magnitude of the hazardous waste problem, we can begin to appreciate that there are no simple solutions available. It has been and continues to be a problem that will require a concerted and sophisticated professional approach to effective prevention, control, and treatment to solve. In this context, then, we intend in this chapter to cover the major broad topics that have to considered in understanding the basic areas of concern and the various elements that have to be asked in developing and implementing an effective hazardous waste management system.

As previously mentioned, one of the difficulties of developing and employing such a system is the very fact that to be effective it must be approached on a systems basis rather than on the basis of examining and considering only discrete elements of the problem. For example, we can consider hazardous wastes either from a single source or in the more generic sense as consisting of a rather simple but nonetheless important linear progression of discrete elements which have been quite reasonably and realistically employed as the basis for structuring the application of our Resource Conservation and Recovery Act (RCRA); these are

- Generation
- Storage
- Treatment
- Disposal
- Transportation

Important and appropriate as these elements are, and if we consider the progression of individual or several wastes from a single source or from a generic category of sources, they do in fact make up the components of a system. In that sense, it would be appropriate, from the perspective of hazardous waste management regardless of categorical source of waste derivation, to consider these elements both individually and collectively — the latter in the systematic sense. Such an approach would mean that one would not only examine existing practices and programs in each of the four basic areas identified but also as each of these areas relates to the other and to the ultimate objectives of either eliminating, reducing, or controlling waste generation, and effective treatment and disposal of those wastes that are generated.

As important as such a system is, it is only one way of examining and establishing a systematic approach to hazardous wastes. The other, which is based more on an examination of managerial options, combined with appropriate consideration of scientific and technological feasibility, involves a systematic analysis of management and control options. In that instance, we would utilize the approach shown in Table 1.

Whichever approach is used — and others of course may be developed with equal utility and validity — the basic point is that, from the design and management viewpoint, these elements are not considered individually but in an interrelated and integrated fashion. Here is one of the areas that has not been adequately examined in the literature and where we feel that this volume and subsequent volumes in this series fulfill a definite need.

With the necessity for integration of activities in mind, it is a relatively simple progression from consideration of these more or less technical considerations to those that are more aptly categorized as managerial oriented factors although there are certainly scientific and technical aspects to be considered when examining these areas. They are important, however, since they provide a structural foundation for an integrated examination of the problem and waste treatment approaches and options either within a given manufacturing, municipal waste treatment plant or nuclear power generation station, or any source of hazardous or, for that matter, nonhazardous waste generation. Therefore, we will be examining all of the following topics in this volume in some detail:

- Management Programs
- Low- and Non-Waste Technology
- Reuse, Recycle, Recovery
- Waste Disposal Options
- Waste Treatment Options
- Waste Characterization

IV. WASTE CHARACTERIZATION

Taking the above list in reverse order, clearly the first logical step in approaching waste treatment is characterizing the wastes themselves and identifying the nature of any particular problems that might be associated with them. This is, of course, also an important factor in determining actual treatment and disposal options. Therefore, it is necessary to determine certain characteristics in order to obtain the necessary information. Essentially this involves two separate determinations:

1. Whether or not a waste should be considered to be hazardous. This is based upon establishing certain criteria such as toxicity, flammability, corrosivity, reactivity, radioactivity, persistence, or other hazards. Obviously the criteria employed are an extremely important factor since waste will be determined to be hazardous or nonhazardous based upon these characteristics and certain established values included in any classification system. These characteristics and values have been established in the U.S. through various regulations issued by the EPA as well as by various state and local authorities. Similarly, many other nations of the world have also established criteria for classification.

2. Characterization of the waste in the sense of determining its chemical composition and the physical-chemical properties of the waste and its various components. Such information as pH, solubility, volatility, physical form, and vapor pressure are all important determinations as are such things as reactivity, thermal stability, decomposition properties, etc. since these data are essential in determining the physical-chemical properties of the waste as well as the opportunities and restrictions in treatment and disposal.

As far as characterization is concerned, there are three fundamental approaches which have been adopted by EPA which are used for classification or differentiation purposes. These are (1) by actually testing the waste and declaring it to be hazardous predicated upon preestablished criteria, (2) listing of substances or classes of substances or characteristics by the agency based upon an agency predefinition of hazardous, and (3) an assumption of hazardous or nonhazardous classification without actual testing. Obviously each of these approaches has its advantages and disadvantages however, from a purely scientific viewpoint, arbitrary classification is at best questionable. It is to be recognized, however, that from an economic viewpoint there may be advantages in preestablished classification systems based on preemptive approaches such as considering all wastes containing chlorinated compounds as being hazardous by definition.

Whatever the approach used, the essential point is still that the waste is categorized as essentially hazardous or nonhazardous. This determination, combined with assessments of physical-chemical properties, volume, etc., is used to establish a foundation for the determination of treatability and treatment. These data could also be useful in identifying opportunities for minimization or elimination of the wastes from the generation viewpoint, although here it is clearly necessary to study the generation processes in their entirety from a chemical processing and engineering viewpoint.

V. WASTE TREATMENT OPTIONS

As we have already seen, classification and characterization are essential not only to distinguish between hazardous vs. nonhazardous waste but in relation to identifying treatment needs and based upon the characterization of the waste treatment options.

At this point, it is beyond our scope in an introductory chapter and volume to deal

Table 2

Underground injection
Landfills
Natural storage areas
    Caves, shale beds, etc.
Secure land fills — burial
Ocean dumping

with each of these waste treatment options in detail; however, as an introduction the basic options available are indicated in the following table.

Waste Treatment Options

Land treatment
Thermal treatment
Incineration
Chemical, physical, and/or biological[a] treatments
    Neutralization
    Precipitation
    Floculaton
Solidification

[a] Including biotechnology-bioengineered organisms.

Clearly the selection or utilization of one or more of these options will be predicated upon not only the physical-chemical considerations we have already mentioned but on such other practical factors as the volume of the wastes themselves, the availability of such treatments in a particular geographic location (which of course has to include consideration of transport and transportability of the wastes), economic factors, and lastly, but perhaps most importantly, the availability of secure dump sites for the ultimate disposal of "hazardous" waste. This latter issue, namely, the identification and establishment of secure hazardous waste dump sites, remains one of the paramount public policy issues to be resolved.

One very important factor in opting for any particular waste treatment approach is the feasibility of actually utilizing the approach. This is, in reality, not only a problem with solid waste but with water and air pollution control technology as well. For example, pollution control or waste treatment techniques which require very large tanks, lagoons, retention basins, or similar availability of substantial land areas are simply not feasible in certain geographic areas. The irony of that is that in fact in many instances, the waste generation or pollution generation source is located in an already highly crowded industrialized region which, because of such geographic restrictions, immediately reduces local treatment options.

VI. WASTE DISPOSAL OPTIONS

As in the case of treatment options, it is essential that waste before and after treatment should be evaluated so that its hazardous or nonhazardous properties are understood in order to properly effect "final" disposal. As in the previous discussion on treatment options, it is beyond our purpose to go into actual disposal techniques in any detail; however, Table 2 lists the major categories of options available. Selection of a particular option in a specific case is again dependent upon the characteristics of the wastes from the viewpoint of either being hazardous or nonhazardous as well as relative to their physical-chemical properties after treatment. It is also dependent upon the same availability, geographic, transport, economic, and related considerations previously identified.

Naturally, associated with these various options are a number of specific technical considerations such as waste solidification and appropriate containerization. Similar considerations of containment techniques, monitoring, and sampling are particularly necessary in the case of disposal of highly toxic or radioactive wastes in secure sites. Moreover, local, geographic, and hydrological considerations also apply (e.g., concerns with leaching into aquifers) which have already created serious potential for, and in some cases actual, contamination of ground water supplies.

VII. REUSE, RECYCLE, RECOVERY

While these three subjects — reuse, recycle, and recovery — are closely related from a technical, engineering, cost, and management viewpoint, they do in fact represent different problems and opportunities. Whether or not they are truly waste treatment techniques (in the sense of treating wastes in order to render them nonhazardous or disposable), they are appropriate to consider within the broad subject of waste management because they do present opportunities for waste reduction or elimination.

As a practical matter, many companies have already applied these techniques and found them to be technologically feasible and economically attractive, although here we must be cautious not to create the impression that they are universally appropriate, technologically feasible, or economical. However, the fact that they may not be appropriate or feasible in a particular situation or perhaps at a given point in time does not distract from the need to consider them. It may be that for a particular plant or waste they do in fact present very substantial opportunities for waste reduction or elimination or in the case of recoverable products, they can even represent an opportunity for selling the recovered material.

It is beyond our purpose for the moment to examine the technology and economics of these approaches, but as mentioned in the introduction to this volume it is our intention to examine these techniques in future volumes.

VIII. LOW- AND NON-WASTE TECHNOLOGY

While this is an approach which, as its name implies, is predicated on reducing or eliminating waste, it differs substantially from reuse, recycle, and recovery techniques to the extent that while reuse, recycle, and recovery approaches may in fact be elements in a low- and non-waste technology system, low- and non-waste technology as a concept is much broader. Here again, we plan in future volumes to address the question of low- and non-waste technology in industrial, municipal, or other waste generation situations in a volume dedicated specifically to this subject. For the present we should merely recognize that low- and non-waste technology encompasses a complete consideration of all elements of waste generation, treatment, storage, and economics as a fundamental element in the planning, development, and manufacturing of a product. This means that in practice one would examine waste generation inclusive of not only solid waste but air and water pollution in every phase of product life from research and development through to processing, production, packaging, and eventually consumer or industrial use.

IX. MANAGEMENT PROGRAMS

From the preceding it is quite apparent that, from the viewpoint of actually managing hazardous waste, there are a number of considerations: scientific, technical, legal,

Table 3

Classification — testing
Contingency plans
Compliance programs
Inventory of processes and wastes
Employee training
Labeling/placarding
Emergency response capability
Reports and recordkeeping capabilities
Internal inspections and audits
Scientific and technical process, treatment,
 storage, disposal assessment
Legal compliance
Planning
Transportation
Storage
Rehabilitation of closed sites

economic, and others involved. The term management itself can apply to different aspects of management — in the sense of controlling wastes and in the more classical sense of administrative concerns. Both are, of course, relevant.

From the viewpoint of managing wastes, we have to be sure that the various elements indicated in Table 3 are at least assessed and either included or excluded in the management program as may be appropriate to a particular waste, facility, or locality.

As can be seen this is quite a lengthy list and more importantly involves a number of different skills and disciplines. Moreover, it calls upon not only scientific and technical but legal, administrative, and financial involvement and participation.

This introduces the second "management" aspect; namely, the more classical concerns of management in relationship to applying management techniques to hazardous wastes and in being certain that the company's overall goals and objectives as well as obligations are met.

As a practical matter it implies that waste management cannot be visualized solely as the responsibility of the engineering department or the production department or whatever may have been historically identified as the responsible unit within the company. Rather, it must be approached on an integrated basis so that the company will be certain that all of the various elements we have identified are brought into the actual hazardous waste management decision process. From the viewpoint of actually accomplishing this objective, it means that management will have to designate specific individuals and units as having the responsibility for waste management programs. These individuals and units must also have senior management support and accessibility to the various other functions in the company such as legal, financial, engineering, production, etc. that would have to be involved in such an integrated hazardous waste management approach. This also implies that in a larger company, hazardous waste management ultimately may be approached through the establishment of a hazardous waste management team. This team would in turn include appropriate representatives from the various functions mentioned.

Underlying all of this is the need to consider hazardous waste management in its totality and the inclusion of waste management in the overall planning process and treating hazardous waste management as just that, a management responsibility.

X. SUMMARY

What we have attempted to accomplish in this chapter is to provide a broad overview and introduction to the identification and classification of hazardous waste as well as the various practical considerations inherent in the development of a hazardous waste program. In subsequent chapters in this volume, we will examine some of the historical background of present regulatory and legal requirements as well as the major applicable statutes themselves. We will also explore not only the obligations of industry but its rights. In addition, we will examine the very important area of economics as it relates to specific elements of hazardous waste management as well as to the broader concerns of such policy issues as the creation of Superfund and taxation. Since, in addition to purely technical considerations, legal aspects such as compliance, liability, emergency response, etc. are so important, they are given extensive coverage in this introductory volume of the hazardous waste management series.

Chapter 2

THE CONSTITUTIONAL FRAMEWORK OF HAZARDOUS WASTE LEGISLATION*

Kenneth G. Bartlett

TABLE OF CONTENTS

* All footnotes appear at end of chapter.

The perils of inadequate hazardous waste management highlighted by such publicized incidents as Times Beach, Missouri have created a startling new social awareness that this subject should be comprehensively regulated. Concern over this subject matter no doubt has been accentuated by a transition within the chemical industry itself. The passage of the Resource Conservation and Recovery Act of 1976 (RCRA), the Comprehensive, Environmental Response, and Liability Act of 1980 (Superfund), advances in resource recovery technology, new concerns about energy consumption, rising costs of land filling of waste, and an explosion of state and federal laws and regulations are but a few of the principal factors instrumental in effecting this present industry-wide transition. The 14th Annual Report of the Council on Environmental Quality (1984) remarks that the management of hazardous wastes has emerged as the "most difficult and controversial" environmental issue of the decade.

Much has been written about the ultimate effectiveness of RCRA and Superfund as regulatory mechanisms to combat the broad evils of improper hazardous waste disposal. Conceptually, the prime objective of RCRA is to close the regulatory gap between toxic substances and the more traditionally regarded types of pollution.[1] Basically, RCRA addresses the handling of hazardous waste at currently (or future) operating facilities, not addressing closed or abandoned sites. An amendment signed into law in November 1984 may now extend RCRA to include subsurface contamination predating RCRA.[2]

RCRA was substantially amended in 1984 with overwhelming support in both Houses of Congress with the act being signed by President Reagan on November 8, 1984.[3] These amendments dramatically address a wide variety of controversial issues including virtual elimination of the small generator exemption,[4] a broad policy against land disposal of wastes,[5] and significantly more stringent limitations concerning permitting[6] and blending of wastes.[7] Under the original RCRA Act, most legitimate recycling of hazardous waste was exempt.[8] Under the 1984 amendments, EPA is now required to establish regulations for users, distributors, and transporters of fuels containing hazardous waste.[9] The Act also requires all producers or distributors of such fuels to include a clearly labeled warning stating that the "fuel contains hazardous waste".[10]

Congress also significantly expanded the citizens suit provisions under RCRA[11] as well as adding a new program broadly regulating underground storage tanks.[12] This program includes requirements for release detection, financial responsibility, prevention, and maintenance of the underground storage of any "regulated substance". This amendment will greatly expand the scope of waste activities within the scope of RCRA.[13]

Procedurally, many of these new provisions depart from traditional statutory amendments in that they are effective immediately.[14] The thrust of these amendments suggests that Congress deliberately shortened EPA's leash in implementing many of these new programs. In terms of their detail and specificity, these amendments clearly indicate that both Houses of Congress were willing to become intimately involved with this subject matter. This willingness has resulted in several innovative procedural schemes, which stand to cause considerable confusion both in the regulatory community and at EPA. For example, several key provisions have been declared by Congress to be self-executing; such broad regulating provisions automatically will become law unless EPA promulgates more specific regulations.[15] This innovative pragmatic approach is certain to lead to protracted litigation as it becomes increasingly unclear as to the status and enforceability of EPA's inaction.

Indeed, it is difficult to understate the impact that these 1984 RCRA amendments will have on increasingly greater populations of the industrial community.

Also discussed below is the Superfund Act, which was enacted in December of

1980.[16] Both Acts were enacted to create a national policy of waste management. As discussed below, one effect of the RCRA Amendments has been to extend the reach of waste and industrial activities regulated by RCRA.[17]

Before focusing on the Constitutional issues, it is instructive to briefly outline the history of these two acts.

I. RCRA: THE ACT ITSELF

RCRA was designed to fill a statutory void left by the Clean Air Act and Clean Water Act. While both these acts require removal of hazardous substances from air or water, both leave unanswered the issue of ultimate deposition of waste.

As with past major federal environmental legislation, a flurry of court challenges were brought testing the validity of these initial RCRA regulations. Section 7006 of the Act limits venue (the place where a lawsuit may be brought) to the U.S. Court of Appeals in the District of Columbia for legal challenges to RCRA regulations. This section also sets forth a time limitation of 90 days "from the date of such promulgation" in which to petition for judicial review of any RCRA regulation.

The administrative convenience and expediency of limiting judicial review of these regulations is obvious. In October 1980, RCRA was amended to expand judicial review of matters relating to individual permits and sites and federal authorization of state plans under Section 3006. These matters may be reviewed in the U.S. Courts of Appeals where the petitioner resides or transacts business. The August 1983 decision of *EDF* v. *EPA* held that attorney's fees are recoverable in successful legal challenges under the so-called "Equal Right to Justice Act" (28 U.S.C. Section 2412) even though RCA did not specifically provide for attorney's fees.

The 1984 amendments will significantly expand the role of private citizens in RCRA actions. Not only may civil penalties now be recovered in citizen suits, but Section 401 now allows nongovernmental plaintiffs to bring "imminent hazard" actions under §7003. Thus, where the Department of Justice has refrained from bringing a Section 7003 action at a particular site, the local citizenry may now file suit independent of whether such a suit is filed by the Department of Justice.

Considering the limited provisions for judicial review of these regulations issued pursuant to RCRA, the flurry of petitions challenging these regulations has been astonishing. The release by EPA of the May 19, 1980 RCRA regulations, for example, spawned the filing of 52 separate lawsuits by various trade associations and corporations. These issues were refined and consolidated under the single case name of *Shell Oil Company* v. *EPA*.[18] The *Shell Oil* case concerned a challenge to the substantive RCRA reflections released by the agency in 1980 implementing Sections 3001, 3002, 3003, and 3004 of RCRA (Phase I regulations).

EPA has attempted to minimize ambiguities within the regulations by announcing in August of 1980 its intention to release a series of Regulatory Interpretation Memoranda (RIMs) and technical amendments to the regulations (TARs). Subsequent to this announcement, EPA has been sporadic in the release of these RIMs.[19]

In addition to these substantive challenges, early litigation also focused the time-frame in which the regulations themselves were being promulgated by the agency in *Illinois* v. *Costle* (no. 78-1689 D.D.C. 1979). Repeated delays in the issuance of these regulations has compromised the effectiveness of the Act. Even a cursory reading of the 1984 amendments reveals that Congress has taken considerable care to insure that any delays in implementing these new amendments by EPA will have a minimal effect on the overall effectiveness of the Act.

Although RCRA's Phase I regulations were released over 2 years late and under court order, they are still a credit to the agency. The time-frame in which EPA had to undertake such a massive regulatory program is noteworthy; comparable regulations under the Clean Air Act (42 U.S.C. Section 7401, *et seq.*) and the Federal Water Pollution Control Act (33 U.S.C. Section 1251, *et seq.*) have had 7 to 10 years for promulgation, which contrasts significantly with less than a 4-year work-up period initially given EPA by Congress under RCRA.

The extraordinarily high costs of industry-wide RCRA compliance has continued to place these regulations under close legal scrutiny. This chapter outlines the legal basis of hazardous waste legislation and the historical limitations associated with legislation of this type.

Integral to this inquiry is a discussion of the various constitutional issues associated with this newest frontier of environmental regulation, especially in terms of the power of the federal government to regulate a particular subject.

On October 21, 1976, the Resource Conservation and Recovery Act was signed into law.[20] Built on the legislative foundation of the Solid Waste Disposal Act of 1965 and the Resource Recovery Act of 1979, RCRA was revolutionary in the degree of regulation it sought to achieve. "A fundamental premise of the statute is that human health and the environment will best be protected by careful management of the transportation, treatment, storage, and disposal of hazardous waste, in accordance with standards developed under the Act."[21]

Within the past decade, federal environmental legislation has increasingly limited the permissible emission and discharge levels of point-source pollutants. The substantial foreclosure of these traditional environmental sinks has enhanced the practice of land storage of toxic waste. Surface run-off and natural percolation have contributed to surface and ground-water contamination as a result of this practice of waste land farming. Many of the sludges, which are regulated under RCRA, are the products of other pollution-abatement activities. Although a substantial portion of improperly stored waste predates both the Clean Air and Federal Water Pollution Control Acts, the cause-effect relationship between the historical thrust of environmental law and the magnitude of the present hazardous waste problem is difficult to overstate. The findings by Congress and the underlying legislative background of RCRA note this historical effect on the present problem.[22]

Several reasons explain why waste management has received less emphasis than air and water environmental regulation. One is the relative ease with which property rights may be defined.[23]

Noted earlier, the 1984 amendments evidence a clear Congressional directive against landfilling and surface impounding of hazardous waste.[24]

> By enacting H.R. 2867 Congress will state its finding that reliance on land disposal, particularly landfill and surface impoundments, should be the least favored method for managing hazardous wastes.[25]

Accompanying this directive is a stringent schedule set by Congress requiring EPA to review all hazardous wastes for possible prohibition from land storage altogether.[26] Congress has also enacted detailed minimum technological requirements for landfill waste facilities requiring Part B RCRA permits. These technical requirements include, for example, that all such landfills have two or more liners, leachate collection, and extensive ground-water monitoring systems.[27]

By November 8, 1986, EPA is required by Section 3004(o) to issue regulations for the installation of liners and lechate collection systems for landfills and surface impoundments.

The 1984 Amendments created Subtitle I creating comprehensive regulation of underground tanks containing petroleum and other regulated substances. This underground tank program is not contemplated to be administered via permit but rather by EPA performance standards to be released by EPA in 1987.

The cost of compliance in retrofitting such a facility with these mandatory requirements will, in practical effect, force other options in the disposal and treatment of RCRA waste.

RCRA contains eight subtitles, with Subtitle C receiving top priority from EPA. Subtitle C, which contains 11 sections numbered consecutively from 3001-3011, establishes the hazardous waste management system. Basically, RCRA seeks to monitor wastes designated under Section 3001 of the Act from creation through their use in commerce, to final disposition at an approved disposal facility. This has been widely referred to as "cradle-to-grave" regulatory system. "The Act defines solid waste broadly, so that included are essentially all substances destined for disposal and not regulated by the Federal Water Pollution Control Act or Atomic Energy Act of 1954." Consistent with the pragmatic definition is EPA's frequent reference to the term "waste stream." This term freely acknowledges that chemical residues and waste products rarely are chemically homogeneous and are often difficult to classify.

The thrust of this definition is the separation of waste into two categories; hazardous waste and other solid wastes. A waste product must be characterized as a solid waste before it can be considered "hazardous waste" subject to regulation under RCRA:

> Hazardous waste is (A) solid waste, or combination of solid wastes, which because of its quantity, concentration, or physical, chemical, or infectious characteristics may (B) pose a substantial present or potential hazard...when improperly treated, stored, transported, or disposed of, or otherwise managed.[28]

RCRA has attempted to close the gap of regulatory control by establishing a comprehensive manifest system under Section 3002(5) of Subtitle C of the Act. Section 3001 of RCRA directed EPA both to establish criteria for identifying hazardous waste and to maintain a generic listing of such waste.

Presently, EPA regulations define Section 3001 wastes using the basic methods, a listing of industrial wastes determined to be hazardous, and a separate method utilizing specific chemical properties. These chemical tests concern properties of toxicity, ignitability, corrosivity, and reactivity. More recently, in February 1984, EPA expanded upon this identification procedure and listed as hazardous waste a group of wastes from a generic category generated during manufacture.

Section 3002(5) directs EPA to promulgate regulations establishing a manifest system "to assure that all such hazardous waste generated is designated for treatment, storage, or disposal in treatment, storage, or disposal facilities." The original regulations issued in November 1980, exempted generators from filing a manifest if their production levels do not exceed 1000 kg/month.[29] This exemption has been construed to apply on a site-by-site basis. A firm may maintain several facilities, but any one site that fails to produce hazardous waste in excess of 1000 kg is therefore exempted. Under the Act, EPA has identified over one hundred wastes as being "acutely hazardous" and has established on exemption production threshold of 1 kg/month if this waste is disposed at a waste facility licensed under RCRA.[30] Section 3002 also requires that hazardous waste generators apply for and obtain an EPA identification number prior to shipping hazardous waste.

The 1984 amendments lowered this exemption from 1000 to 100 kg/month.[31] This reduction will have an immediate effect of bringing literally millions of previously unregulated activities within and under the provisions and sanctions of RCRA.

Under the 1984 Amendments, Section 3001(d) directs that EPA create regulations for those generators of hazardous waste of 100 kg/month or more. In accomplishing this, EPA has designed generators in the 100 to 1000 kg/month as a special category of "large-quantity generators" and now considers a "small quantity generator" as those creating less than 100 kg of hazardous waste per month.

A survey released by EPA in March 1985 revealed that more than 50% of the companies falling within the 100 to 1000 kg range of monthly generation of hazardous waste were in five categories: dry cleaning, laundry, vehicle maintenance, metal manufacturing and finishing, and lastly, printing. EPA has predicted that the lowering of this generation threshold will regulate more than 800,000 metric tons of previously unregulated hazardous waste.

The preparation of a manifest is the principal responsibility of the hazardous waste generator. Generally, a separate manifest for each site or location of hazardous waste activity must be submitted to the EPA regional office serving the area in which the activity is located. This document requires disclosure of the legal owner of the installation, identification, and description of the waste product and the EPA identification number of all parties expected to transport and receive the waste.[32]

Similar to a bill of lading, the signed manifest is the waste's ticket in commerce. After disclosing the above information, the generator of the waste must sign the certification on the original manifest. Ideally, this signed document then accompanies the waste as it is transported to a licensed treatment, storage, or disposal facility, as regulated under Section 3005 of RCRA. The generator's receipt of the waste facility operator's copy of the manifest completes the record-keeping process.

Sections 3004 and 3005 of the Act have directed EPA to issue regulations establishing standards and permit procedures for facilities treating, disposing, or storing hazardous waste. All hazardous waste management facilities in existence under the rule which came into effect in November 1980 (including those that failed to qualify for interim states) are subject to the interim states standards issued under RCRA.

A report issued by the General Accounting Office in June 1984 has pointed out how far behind this permitting process under RCRA actually is. By March 31, 1984, both the states and EPA combined had only issued 132 permits for such hazardous waste facilities. EPA had projected to issue approximately 950 such permits to RCRA regulated facilities by September 1983.

Attempts have been made with these new amendments to speed up the entire permitting process. Whether this will be accomplished, of course, remains to be seen. For example, Section 213 sets specific time limitations for submitting full permit applications for land disposal facilities — 12 months from the date signed into law and requires EPA to issue the final permit within 4 years. Similar time limitations also were enacted for incinerator facilities seeking permits.

EPA has issued Consolidated Permit Regulations (CPRs) designed to expedite the permit procedures under four EPA administered programs.[33] These CPRs were issued in part in response to former President Carter's Executive Order 12044 and were subsequently challenged for alleged procedural and other deficiencies in *NRDC* v. *EPA*, and consolidated cases.[34] A settlement agreement in this matter was reached in December 1981, touching upon a variety of RCRA issues, including partial modification of the permitting process.

In addition to the Hazardous Waste Management program under Subtitle C of RCRA, the CPRs include: (1) the Underground Injection Control (UIC) permit program under the Safe Drinking Water Act;[35] (2) the National Pollutant Discharge Elimination System (NPDES) portion of the Clean Water Act; (3) Section 404 Dredge and Fill permit programs also under the Clean Water Act; and finally (4) some procedural requirements for the Prevention of Significant Deterioration (PSD) program under the Clean Water Act.

A. Legislative History of RCRA

The original legislative history of RCRA was sparse, considering the ends this legislation sought to achieve. It was passed hastily and without extensive discussion. The original complete legislative history of the Act consisted of one House report[36] and two Senate reports.[37]

While it is perhaps tempting to blame EPA for the shortcomings and misguided efforts of RCRA, the legislative vacuum that EPA began with in 1976 should be noted. In late 1978, EPA itself commented on the utter lack of legislative history. "There is not even a conference report. Thus, EPA has been forced to make several initial policy decisions without the assistance of clear Congressional direction."[38]

It is ironic to note the extent to which Congress has taken a 180° turn with respect to its oversight of EPA in enacting the 1984 amendments. In 1984, a Conference Committee was convened to resolve differences in the proposed amendments.[39]

Legislative history is also often helpful in determining whether Congress intended the federal Act to preempt state legislation. Preemption, a doctrine that reaches the very structure of our constitutional system, requires invalidation of inconsistent state legislation where Congress has passed national legislation that addresses the same subject matter.

Analysis of the preemption issue focuses largely on statutory interpretation. For this reason, courts will often examine legislative history to determine whether Congress actually intended the federal law to preempt state law. Such an examination is necessary when a particular federal Act, such as RCRA, does not specifically direct federal preemption of state legislation. Judicial inquiry into the purpose of the legislation may assist in determining whether preemptive intent may be inferred. As noted later, portions of RCRA's legislative history contain specific language indicating that drafters of this legislation believed federal preemption of state solid waste legislation was unwarranted and undesirable.

B. The State Role

Within Subtitle C, Congress indicated an active state role in the regulatory process. RCRA is characteristic of this newer type of environmental legislation in which individual states will implement and maintain the day-to-day regulatory process, pursuant to express federal standards. An amendment to Section 3009 passed in October 1980 specifically authorized states or political subdivisions to enact regulations that are more stringent than those imposed federally.

Section 3012 was also added in the October 1980 amendments to RCRA, making it incumbent on each state to "submit to the Administrator an inventory describing the location of each site within such state at which hazardous waste has at any time been stored or disposed of."

The partnership of state and federal efforts was outlined in the original objectives of the Act.[40] It is clear from the original legislative history that states are the preferred level of government for the implementation of this program. Section 3006 of Subtitle C specifically directed the EPA administrator to promulgate guidelines to assist states in the development of their individual programs.

As with many of the federal programs of this type, federal funding is provided to assist the states in administering these state programs. EPA has indicated recently that it may begin to hold back funding to those states which have unsatisfactory enforcement records.

It was originally contemplated that virtually all of the states would ultimately receive final authorization to operate their own RCRA program with EPA performing only an oversight function. This delegatory scheme, similar to that of the Clean Air Act, is broadly described as "Cooperative Federalism" whereby the day-to-day functional operation of the program itself is administered by the states. While the advantages of local administration are self-evident, the process of shifting the functional operation to the state level creates substantial practical logistical problems. In applying for a permit application, for example, any applicant seeking a permit must verify whether the state has assumed legal responsibility for the particular aspect that the permit addresses. This difficulty is heightened in two respects. First, the application process often spans a period of time during which there may be several shifts in administration of the program. Secondly, the interim state authorization may be granted to a limited subject area (e.g., tanks and containers, but not land disposal facilities), thus making it more difficult than first apparent.

The overwhelming majority of the states presently have received interim authorization for Phase I implementation of RCRA. This authorization pertains to generator standards and manifest requirements. Of the few states remaining, many have executed cooperative agreements allowing the states to administer particular portions of the federal Act.

It is important to note that a state may receive interim authorization if its program is the "substantial equivalent" to the one administered by EPA.[41] This section also prohibits from giving a state RCRA program final authorization unless (1) it is "equivalent to" the federal program, (2) it is "consistent with" the federal program *and other state programs,* and (3) it provides for adequate enforcement. The 1984 amendments also track the "substantial equivalent" language.[42]

The 1984 RCRA Amendments remove interim authority from any state that lacks final authority to administer a RCRA plan by January 31, 1986 and requires that operation to revert back to EPA. Approval of "final" state plans has been tedious and slow; Delaware received its approval in December 1983, Mississippi in June 1984. As of December 1984, only ten states had even submitted applications for final authorization.

As mentioned before, an October 1980 amendment provided that the state programs may contain provisions more stringent than those received federally. Some states, based on this amendment, for example, have enacted regulations requiring the filing of a manifest for each load, to allow for more careful monitoring of hazardous waste in that state. Present federal regulations only require reporting to EPA if the manifest cannot be reconciled with the generator within 15 days of receiving the waste. Other states such as Vermont and Rhode Island have expanded their interim authorization plans to include radioactive and infectious wastes. These two states, in addition to California and Connecticut, have opted not to establish a small generator exemption at all.

The patchwork of individual state requirements as illustrated before has created substantial practical problems for multistate companies interested in developing a uniform national program which will satisfy all state regulatory programs.

Although RCRA provides a mechanism whereby states may operate their own program in lieu of the federal counterpart, implementation of this scheme at the state level continues to be difficult for several reasons. First, the state authorization procedure itself under which a state assumes a primary role is substantially more complicated than with other federal Acts, particularly since interim authorization may be granted to portions of federal regulatory schemes. This observation is attributable largely to the provision within RCRA for interim state authorization of the respective state plans. This wrinkle of complexity is the by-product of the relatively short time period in which this regulatory effort has been launched.

Second, the potential for conflict between the respective state plans is much greater than with other similar federal-state regulatory schemes, such as the Clean Air Act or the Safe Drinking Water Act. The very nature of the subject creates this heightened potential for conflict among state plans.

Under this scheme, it was contemplated that complete state RCRA applications for final authority were to be submitted to EPA by July 31, 1984. These final state plans were to confirm with the federal statute by that July 1984 date.

Concern has recently been expressed by EPA that some state RCRA programs may revert back to EPA because of inadequate levels of personnel and state funding. Alabama, for example, relinquished control of its interim state plan on August 1, 1984.

Section 228 of the 1984 amendments requires that "any requirement or prohibition...shall take effect in each state having an interim or finally authorized state program on the same date as such requirement takes effect in other states." This section further provides that if evidence is presented to EPA showing that the requirements of the state program are "substantially equivalent" to the federal scheme, the state may operate that portion of the state plan. While the operation of this provision is, at the present time, far from clear, it appears that Congress wants stricter control of the state programs. For example, unless authorization is obtained by the particular state on or before the effective date, EPA will administer that particular requirement.

This retraction of sorts from the state plans is an attempt to minimize the patchwork of individual state plans that were developing under RCRA.

II. SUPERFUND

Fundamentally, "Superfund" is concerned with hazardous waste contamination from spills and abandoned dumpsites when there is no other party taking prompt action to clean up the affected areas.[43] The funding provisions for CERCLA, which provided $1.6 billion over a 5-year period, expired on September 30, 1985. It has come up for reauthorization and at the time of this writing, the House had passed a bill (HR2005) increasing the Superfund more than sixfold by sharply increasing its tax on basic chemicals and oil. It is presently contemplated that it will be reauthorized for a period of 5 years and will not be signed before the spring or summer of 1986. This Act has been the subject of intense public debate and scrutiny pertaining to political overtones in enforcement of this Act.

Essentially, superfund requires any person in charge of a vessel or facility where a hazardous substance is located to notify the National Response Center as soon as he has knowledge of a release or potential release of the substance. Any individual failing to comply is subject to a fine, imprisonment, or both.

While the scope and nature of generator liability has been discussed in considerable detail in subsequent chapters, it may be stated in general terms that 107(a) of CERCLA has been interpreted broadly. While far from clear, the section imposing liability states that basically, a "release" is defined broadly, including any spilling or leaking, pumping or pouring, or dumping or disposing of hazardous materials. There are five types of discharges excluded from the Act's record-keeping and reporting requirements: (1) releases resulting in exposures to persons solely in the workplace if the employees affected are able to assert a claim against the employer; (2) certain engine exhausts; (3) certain releases of nuclear materials; (4) the normal application of fertilizers; and (5) any federally permitted release under a federal or state permit system (47 U.S.C. 9607(a)).

When a release does occur, or when there is a substantial likelihood of a release into the environment, the President, acting through EPA, is authorized to take remedial action, consistent with the National Contingency Plan to remove the hazardous substance, pollutant, or contaminant. The National priorities list identifies areas eligible for Superfund response under the act. This list has steadily grown from 115 sites identified in October 1981 to more than 780 sites in October 1984.

The owner or operator of any vessel or facility from which a hazardous substance is discharged is strictly liable to the U.S. for the actual costs incurred by the government for the removal or clean-up of such materials. These actual costs include expenses to restore the environment to its natural state.

In extraordinary instances, it is possible for an owner or operator to avoid liability for these clean-up expenses. Although there is strict liability, Section 107(b) allows a defendant to assert as a defense that the damages complained of resulted from acts of war, acts of God, or acts of independent third parties where the defendant establishes that he has exercised "due care" and took precautions against the third party's foreseeable acts of omissions. This so-called third party defense is inapplicable to persons with whom a generator directly or indirectly had or has a contractual relationship.

A federal district court in Missouri held that a restitution action brought under Section 107 is purely equitable in nature and therefore, the defendants were not entitled to a trial by jury.[44]

In order to expedite clean-up, funding has been provided by the Act. A Hazardous Substance Response Trust Fund has been created, financed jointly by industry and the federal government, and monies in the fund are available to the heads of federal departments, agencies, and instrumentalities who are involved in removal and clean-up operations. Of course, funds will not be expended if the agency determines that the owner of the vessel or facility will remove the materials and restore the environmental balance.

Another distinct feature of Superfund is that it allows for federal and state cooperation in cleaning up affected areas which have been abandoned. The state, by way of agreement, may assume responsibility for the initial clean-up. Superfund will then reimburse the state for 90% of the removal costs requiring the state to pay the subsequent 10% and any future maintenance of the site.

Basically, sources of revenue for the fund are taxes on crude oil, impure petroleum products, and certain basic industrial products; yearly appropriations from the general fund; amounts reimbursed from owners, operators, or transporters on behalf of the fund; monies collected under the oil and hazardous substances liability provision of the Clean Water Act; and penalties and punitive damages provided for in the Act itself.

Lastly, the Post-Closure Liability Trust Fund has been created by the Act. This fund assumes the liability of owners or operators of hazardous waste disposal facilities when the owner has complied with the requirements of RCRA and the disposal site has been closed for 5 years with no apparent likelihood of releases or damage to the surrounding areas.

In August 1984, the House passed a measure increasing sixfold (to $10 billion) the monies available under Superfund to clean up abandoned hazardous waste dumps.

Removed by an amendment was a section of the bill which would have provided a federal cause of action for toxic tort victims.

Since the enactment of Superfund in December 1980, a relatively clear line existed between activities that were subject to Superfund and those activities falling under RCRA. This distinction between these two acts has been muddled by Section 206(U), now requiring "corrective action for all releases of hazardous waste...at a treatment, storage, or disposal facility seeking a permit under this subtitle, *regardless of the time* at which the waste was placed in such unit." This amendment now creates RCRA jurisdiction and liability to contamination sites where previously such a site would only be subject to Superfund.

With this background, we now turn to discussion of the Constitutional framework which authorizes Congress to institute this legislation.

III. CONSTITUTIONAL FRAMEWORK

A. Federalism

An analysis of hazardous waste legislation should begin with an examination of the basic relationship between state and federal governments. This relationship is defined by the principals of federalism and the commerce clause.

The Congressional power most frequently invoked in hazardous waste legislation such as RCRA and Superfund is the power to regulate interstate commerce. Article I, Section 8, Clause 3 of the U.S. Constitution confers upon Congress the power "to regulate commerce with foreign Nations, and among the several States, and with the Indian tribes."

Through Congress, the federal government has extremely broad discretion to enact legislation with objectives directed at correcting economic imbalances. One legitimate facet of this power to regulate interstate commerce is the reasonable regulation and control of hazardous waste.[46]

Although aimed at environmental concerns, this type of legislation is characterized as a type of "economic" legislation and therefore reachable by the commerce clause. As with air and water pollution, Congress, in enacting these acts, concluded that federal regulation of hazardous wastes was necessary because of market failures and inadequate state regulation.

However, Congress has not always had authority over this area. Under the original Articles of Confederation, each state was a sovereign unto itself, and, as such, had virtually an unlimited right to tax and surcharge goods and wares passing through its borders. This piecemeal erection of ad hoc revenue-raising measures by the individual state significantly impeded the flow of commerce among the states.

Such was the situation with interstate commerce before the Constitution Convention called in Philadelphia in 1787; Congress could act only in an advisory capacity. It was in this context that those delegates meeting at the convention opted to remove from the individual state sovereignties this plenary right to regulate wholly commerce within their borders and create the federal authority to benefit all the states. The often quoted language of *Parker* v. *Brown* expands this concept of independent sovereignties:

> The governments of the states are sovereign within their territories save only as they are subject to the prohibitions of the Constitution as their action in some measure conflicts with powers delegated to the National Government, or with Congressional legislation enacted in the exercise of those powers.[47]

The tenth amendment has been given a limited reading; the Court's retraction from the position taken in *Usery* in 1976 to that in *Garcia* concerning the policing of the Tenth Amendment has been previously commented upon.

The authority of the commerce clause has been relied on most often to justify federal environmental legislation.[48] This right to regulate interstate commerce is greatly enhanced by the "necessary and proper" clause.[49] Thus, in asserting the validity of federal legislation, Congress has been given broad discretion in regulating interstate commerce through the combination of the two federal grants of power discussed above and a series of broadly construed judicial decisions.[50] As Professor Rosenthal has observed:

> Thus, anything which may be properly characterized as "interstate commerce" may be regulated, or indeed forbidden by Congress. But that is not all. The power to regulate or protect interstate commerce extends to include the power to control those things which are not commerce or are purely intrastate commerce, provided that such control is reasonably deemed by Congress to be of assistance in the regulation or fostering of interstate commerce itself.[51]

Historically it was the duty of the courts in passing upon the constitutionality of the Act to determine whether the exercise of police power is necessary for the public good.

In determining the validity of such an exercise of a state's police power, the courts primarily focused on the real character of the Act and on the legislative end to be accomplished, rather than its title or any declared purpose. Thus, in addressing this delicate issue of whether a state Act overly interferes with property or liberty rights secured by the federal constitution, the court's inquiry focused on the practical operation and effect of the Act, not on how it is characterized. An incidental restriction pursuant to an exercise of police power is valid if it bears a real and substantial relation to the health, safety, morals, or general welfare of the public, and if it is not unreasonably or arbitrary.

The court's role as a watchdog over the Tenth Amendment has significantly lessened. In February 1985, the Supreme Court issued its decision in *Garcia* v. *San Antonio Metropolitan Transit Authority* (SAMTA), giving a very limited interpretation to the Tenth Amendment.[51a] In ruling that the Congress and not the judiciary is better suited to deal with the Tenth Amendment's limitations on national power, this 1985 decision reveals that the present Supreme Court has given the Tenth Amendment a very limited interpretation. *Garcia* represents a rather significant departure in the Court's perception of its role in policing the balance of power between the power of the state via the Tenth Amendment and the national powers.[51b]

While it is beyond our present scope to explore this matter in greater detail, it is certainly appropriate to comment on the degree to which this entire subject is influenced by political forces. The observation, for example, that there is a drift nationally to more conservative ideologies should not be underestimated in its impact.

Thus, two separate sources may be identified authorizing federal hazardous waste legislation, these being the affirmative grant of power to Congress to regulate commerce via the "Commerce Clause" and secondly, the principals of federal supremacy "federalism" arising from the supremacy clause of the U.S. Constitution. Before focusing further on these two concepts, it is worthy to discuss briefly the related Constitutional principal of due process.

B. Due Process and Equal Protection

The due process and equal protection clauses of the federal and state constitutions guarantee basic fairness of any legislative scheme. The 5th and 14th Amendments both contain a due process clause mandating that federal and state legislation be reasonably related to the legislative purpose sought.

The equal protection clause of the 14th Amendment prohibits states legislatures from enacting laws that overly discriminate against certain classes of persons. The 5th Amendment has no equal protection clause guaranteeing that Congress enact its law with a rationally based classification scheme. However, the equal protection clause of the 14th Amendment has been read into the 5th Amendment, thus mandating that Congress may not discriminate unreasonably in the legislative classifications it creates. The 14th Amendment states, "No State shall deprive any person of life, liberty, or property, without due process of law; nor deny to any person within its jurisdiction the equal protection of the laws." Case law, however, reveals the courts deferential posture where Congress is framing economic legislation, on the other hand, to satisfy a substantive due process claim, "the challenged legislation must have a legitimate public purpose based on promotion of the public welfare, health, or safety and be rationally related to the accomplishment of (that) legitimate state purpose (*Alladin's Castle, Inc.* v. *City of Mesquite,* 630 F. 2d. 1029, 1039 [5th Cir. 1980]). Because of the relative ease with which a "rational purpose" can be found, procedural due process issues have provided more fruitful litigation.

A recent procedural due process argument was successfully made in *Dirt, Inc.* v.

Mobile County Commission, 739 F. 2d. 1562 (11th Cir. 1984), where the plaintiff landfill operator challenged the actions of the defendant commission in failing to notify him that this landfill permit would be considered at a scheduled meeting. Although the 11th Circuit upheld Alabama's waste management statute as being neither overly vague or beyond the permissible scope of the state's police power, the court held that the landfill operator's procedural due process rights were violated in the commission failing to notify him of the hearing.

Remedial economic legislative schemes, such as RCRA and Superfund, historically have been given wide latitude by the courts. The effect of this judicial posture in adopting a "minimum rationality" standard of judicial review is that the legislation generally survives challenges alleging violations of due process or a denial of equal protection.

It is therefore unlikely that either RCRA or Superfund will be seriously challenged by this "minimum scrutiny" standard of judicial inquiry since the courts have traditionally demanded little evidence from the legislature to uphold the Constitutionality of this type of legislation. "It is by now well established that (such) legislative acts (of economic regulation) come to this court with a presumption of constitutionality, and that the burden is on one complaining of a due process violation to establish that the legislature has acted in an arbitrary or irrational way.[52]

The leading case applying this "minimum rationality" standard is *Williamson* v. *Lee Optical of Oklahoma*[53] in which the Supreme Court, in a unanimous opinion written by Justice Douglas, held that certain provisions of an Oklahoma statute regulating visual care were not unconstitutional. The deferential posture of the Court has been adopted in subsequent decisions rendered in constitutional challenges to environmental legislation. The following language in *Lee Optical* illustrates this cursory standard of judicial review applied to the evaluation of the appropriateness of a legislature's particular approach:

> But the law need not be in every respect logically consistent with its aims to be constitutional. It is enough that there is an evil at hand for correction and that it might be thought that the particular legislative measures was a rational way to correct it.

> The day is gone when this Court uses the Due Process Clause of the Fourteenth Amendment to strike down state laws, regulatory of business and industrial conditions, because they may be unwise, improvident, or out of harmony with a particular school of thought.[54]

Closer to the subject matter at hand is the rejection of the equal protection argument advanced by several interstate bottlers in *American Can Co.* v. *Oregon Liquor Control Commission.*[55] In *American Can,* the bottlers unsuccessfully asserted that the Oregon statute violated the equal protection clause of the 14th Amendment in singling out metal and glass beer and soft drink containers without regulating containers for other beverages or substances. The court was satisfied with the mere rationality of this recycling scheme for these selected containers and barely addressed the equal protection issue.

Also significant is that the Oregon court did not engage in a "balancing of interests" test characteristic of a commerce clause analysis as discussed below. This decision has been summarized to state the following principal of law:

> The rationale of both lower courts was rather facile; since the law did not regulate transportation per se, had no discriminatory intent, gave no special exemptions to resident industries, and served a legitimate state purpose, it did not contravene the commerce clause. Both courts saw no need to engage in a balancing of interests, since the objection of the regulation was not transportation.[56]

Therefore, challenges based on equal protection grounds appear unlikely in view of a legal posture characteristic of *American Can.*

A common theme noted throughout this discussion has been the importance of legislative history. As with preemption, courts look to the underlying legislative history when confronted with equal protection challenges. Although broad latitude has been accorded the legislation, as noted above, still greater judicial deference will occur when there is clear legislative intent.

C. Administrative Review and Ripeness

The emphasis of environmental law has shifted subtly, but materially since its advent approximately 15 years ago, and the enactment of the National Environmental Policy Act (NEPA) in 1970.[57] Since the early NEPA cases, which focused primarily on procedure,[58] the percentage of litigation concerning review of agency regulations has increased. It is estimated that two thirds to three quarters of all environmental cases now involve administrative or agency review of regulations, at the state, local, or federal level.

Judicial review of agency regulations and decisions is statutory and is set forth in Administrative Procedure Act (APA).[59] Section 7006 of RCRA provides that (with minor exception) all judicial review of agency regulations will be in accordance with the APA. The scope of review is codified in Section 706 of the APA.

The APA provides for two types of procedural rulemaking: formal rulemaking pursuant to Sections 556 and 557 where the particular Parties are before the agency and proceed in essentially an adjudicative manner. This formal rulemaking procedure is only utilized if the agency's organic statute requires rules to be promulgated "on the record...after...hearing." These above sections of formal rulemaking also apply to adjudication proceedings before an agency.

In all other cases where such a formal procedure is not required by the Agency's Organic Act, procedural rulemaking may be more informal and typically is one of notice and comment. After notice of the proposed rule is published in the *Federal Register* and interested persons are given an opportunity to comment, the agency may allow an opportunity for oral presentation.

The APA establishes an "arbitrary and capricious" standard of judicial review for "informal notice and comment" rulemaking characteristic of EPA, as authorized by Section 706 (2) of the APA. As set forth in a footnote, such informal rules may be set aside if they are "arbitrary, capricious, an abuse of discretion, or otherwise not in accordance with law."[60]

Despite this mandated standard of judicial review, willingness of lower courts to review administrative rulemaking procedures has been diminished by the Supreme Court's posture in *Vermont Yankee Nuclear Power Corp.* v. *National Resources Defense Council.*[61] In *Vermont Yankee,* the court specifically held that with respect to informal rulemaking, courts may not require administrative agencies to provide rulemaking procedures in excess of those authorized by Section 553. While this holding is quite narrow and did not even concern the EPA, the Court's strong language nonetheless directs a distinct trend away from lower court judicial intervention with informal agency rulemaking procedure. Faced with "judicial intervention run riot", which bordered on the "Kafka-esque", the Court further stated, "We have also made it clear that the role of a court in reviewing the sufficiency of an agency's consideration of environmental factors is a limited one, limited both by the time at which the decision was made and by the statute mandating review."[62]

The tone in terms of judicial review of the *Vermont Yankee* decision understandably has affected other courts instrumental in the review procedure. The following language is characteristic of the present judicial reluctance to intervene (significantly, venue for review of RCRA regulations is the District of Columbia Court of Appeals):

> In short, we are willing to entrust the agency with wide-ranging regulatory discretion, and even, to a lesser extent with an interpretive discretion vis-á-vis its statutory mandage, so long as we are assured that its promulgation process as a whole and in each of its major aspects provides a degree of public awareness, understanding, and participation commensurate with the complexity and intrusiveness of the resulting regulations.[63]

One of the earliest cases file under RCRA, *Citizens For a Better Environment* v. *Costle*, presents another limitation of judicial intervention into administrative matters; that of the concept of ripeness.[64] The Appellants (CBE) asked the Court of Appeals in the D.C. circuit to require the Agency's Administrator to provide standards in RCRA regulations for determining whether sewage sludge qualifies as a hazardous waste under the Act. The Court declined to decide this question, concluding that the case is "not ripe for consideration at this time".[65]

Accordingly, it appears that EPA will be accorded significant discretion in responding to its statutory mandate.

IV. PREEMPTION

A successful attack on state legislation based on preemption grounds requires a judicial determination that "Congress intended to occupy the specific area of state legislation", although the factors considered and the willingness of the court to ascertain "legislative intent" have not been wholly consistent. If a state law or regulation conflicts with a federal statute, the state enactment clearly is subject to federal preemption.[66]

Sometimes, Congress expressly states that state legislation is preempted. It is more often the case that Congress is silent on this issue, leaving the particular federal agency to argue that Congress intended that the particular legislation preempt state legislation by "implication". In the latter case, it becomes the duty of the courts to determine the extent, if at all, Congress intended that area of state legislation preempted. Federal law will be viewed as preempting state legislation on that subject by implication when there is an irreconcilable conflict between the two laws, or when Congress, through statutory language or legislative history, has expressed a "clear and manifest" purpose to "occupy the field" addressed by federal legislation. Factors relevant to such a judicial finding — that Congress intended the federal law to preempt a subject matter — have been summarized:

> Especially relevant for preemption purposes are indications that national uniformity was a major concern. When Federal action is inspired by a desire to avoid multiple and conflicting state regulation, or to circumvent the parochial attitude of local authorities, the context strongly suggests that the states should not be allowed to continue to govern matters subject to Federal regulations.[67]

The recent Supreme Court case of *Silkwood* v. *Kerr-McGee Corp.* presented an opportunity for analysis and discussion of this doctrine. In *Silkwood*, the high court upheld a state provision for punitive damages for injuries caused by nuclear hazards even though the Nuclear Regulatory Commission (NRC) was vested with exclusive regulatory authority of the safety aspects of nuclear power. The injury involved the well-publicized contamination of plutonium of Karen Silkwood at the Kerr-McGee plant.

Implicitly upholding the state interests and policy arguments, by upholding the jury awards the court acknowledged the tension flowing between the Federal authority to regulate on one hand and the interests of the state in promoting safety on the other.

> As we recently observed . . . state law can be preempted in either of two general ways. If Congress evidences an intent to occupy a given field, any state law falling within that field is

preempted . . . If Congress has not entirely displaced state regulation over the matter in question, state law is still preempted to the extent it actually conflicts with federal law, that is, when it is impossible to comply with both state and federal law . . . or where the state law stands as an obstacle to the accomplishment of the full purposes and objectives of Congress . . .

Silkwood v. *Kerr-McGee Corp.,* 104 S.Ct. 615(1984)

Another case of interest is the Seventh Circuit Court of Appeals case of *Brown* v. *Kerr-McGee Chemical Corp.* (84-1294) (July 18, 1985). In *Brown,* the Court found the state's attempt to order removal of "byproduct radioactive material" to be preempted by the fact that the NRC has "exclusive authority to regulate the radiation hazards of the byproduct material". While affirming the district court, the Court noted that, if the state injunction was allowed to stand, it would be "an obstacle to the accomplishment of the full purposes and objectives of federal regulation of radiation hazards". (15 EIR 20690) citing the 1984 Supreme Court case of *Silkwood.*

In *City of Philadelphia* v. *New Jersey,*[68] the U.S. Supreme Court found that Congress, in its enactment of RCRA, had not preempted state law in the field of solid waste. This New Jersey statute and the regulations issued pursuant to it were held to violate the commerce clause.[69] This New Jersey legislative scheme, sought to prohibit the importation into New Jersey of most solid and liquid waste. Despite reversal, the high court agreed with the New Jersey Supreme Court in finding that there had been no federal preemption of state law: "We agree with the New Jersey court that the state law has not been preempted by Federal legislation." In a footnote, the Court expressed its rationale in reaching this decision:

> From our review of this Federal legislation, we find no "clear and manifest purpose of Congress"...to pre-empt the entire fields of interstate waste management of transportation either by express statutory command...or by implicit legislative design...In short, we agree with the New Jersey Supreme Court that ch. 363 can be enforced consistently with program goals and the respective federal-state roles intended by Congress when it enacted the federal legislation.

The Court's abbreviated discussion and summary dismissal of the preemption issue is less than consistent with past Court decisions.[70] Specifically, "preemption may exist when the state legislation is incompatible with the objectives of Congress," without necessarily showing a "clear and manifest" purpose of Congress to preempt the state legislation.

This is reflected in the following language of *Jones* v. *Rath,*[71] which is characteristic of the inquiry made by the Court in ascertaining conflicting objectives requiring federal preemption of state legislation. "The criterion for determining whether state and federal laws are so inconsistent that the state law must give way is firmly established in our decisions. Our task is 'to determine whether under the circumstances of this particular case, [the State's] law stands as an obstacle to the accompaniment and execution of the full purpose and objectives of Congress'."[72]

This unwarranted dismissal of the preempted issue has been attributed to the continuing inability of the Burger Court to arrive at a consensus on federalism and state sovereignty issues.[73] Despite this division, preemption invalidations of state legislation are doctrinally favored because federal power stems directly from the supremacy clause and therefore avoids the vague policy arguments often required in weighing the state legislation's burden on interstate commerce.

Notwithstanding the summary dismissal of this issue in the *City of Philadelphia* v. *New Jersey* case, the statute of the preemption doctrine merits further discussion in terms of possible preemptive capability of future challenges to state legislation. Mentioned above, the leading case involving a conflict in the objectives of the state and federal legislation sufficient to require preemption is *Jones* v. *Rath.*[74] "Preemption analysis requires us to consider the relationship of state and Federal laws as they are interpreted and applied, not merely as they are written.''[75] This distinction calls for further examination of the legislative history of RCRA.

As noted previously, the original legislative history of the Act consisted of two Senate reports and one House report. The House report reveals ''interstate cooperation'' to be a ''basic goal'' of the statute. The following text from the House report reveals that Congress declined to preempt the states from enacting legislation concurrently.

> It is the Committee's intention that federal assistance should be an incentive for state and local authorities to act to solve the discarded materials problem. At this time federal preemption of this problem is undesirable, inefficient, and damaging to local incentive.[76]

This legislative insight reveals a high priority at the time RCRA was enacted in 1976 in mandating an integrated, cooperative effort in the states' management of this subject. As discussed earlier, Section 3006 sets forth the various requirements that states may apply to carry out their individual programs in lieu of the EPA.

RCRA allows authorization of a state plan under Section 3006 assuming the plan is not less stringent than the federal standards.[77] In practical effect, this means that every approved state plan will regulate at least those waste streams contained in the Section 3001 federal listing. This ministerial directive provides the individual states with a regulatory floor to which all state plans seeking approval must conform. Thus, if a state plan has been approved, its objectives have been found to be consistent with the federal Act. Section 4007 also authorizes the administrator to withdraw his or her approval of the state plan after there is an opportunity for a public hearing.[78] The extent to which a state will be permitted to supplement the minimum federal standards rests with the regional EPA administrator.

With respect to Superfund, a number of states have passed statutes which have provided funds to clean-up sites. These funds, produced by imposing taxes and fees on generators or disposal facilities, require industry to pay for the clean-up. Superfund, however, has a preemption clause which places doubt as to the legal liability of these statutes. Subsection 114(c) of CERCLA states that ''[e]xcept as provided in this Act, no person may be required to contribute to any fund, the purpose of which is to pay compensation for claims for any costs of response or damage of claims which may be compensated under this title.''

A case in point is the recent case of *Rollins Environmental Services* v. *Parish of St. James,* a November 1985 decision by the Fifth Circuit Court of Appeals. While the preemption subject included PCBs and the Toxic Substances Control Act (15 U.S.C. 2601 et seq.) it is clear that this federal decision will have direct and far-reaching effects on future attempts by municipalities in attempts at banning waste disposal plants within their borders. The challanged ordinance defined ''areas of special concern'', which included schools, day care centers, nursing homes, grain elevators, etc., so as to create, in effect, virtually a total prohibition against the operation of a commercial solvent cleaning business within the perimeter of the municipality. The following language is instructive as to how the Court used both preemption and commerce clause principals to invalidate the ordinance:

> At the very least, an exercise of legislative rulemaking authority must be a reasonable means of attaining legitimate governmental objectives . . . Here, of course, the question is not so much whether the challenged Ordinance is rationally related to legitimate objectives, but whether it trenches impermissibly upon a field preempted by Congress. Nevertheless, the two analyses are related. Insofar as Ordinance 85-1 amounts to an outright ban of prohibition of appellant's PCB disposal activities, it has the illegitimate objective of regulating a field preempted by Congress. Alternatively, even viewed as a ''commercial solvent cleaning'' regulation, the Ordinance would be an unreasonably burdensome and restrictive means of attaining that end, and would thus be in violation of the Commerce Clause.

While, again this case included TOSCA, it is clear that the preemption and commerce clause arguments are applicable to RCRA as well. The phrase "may be compensated", if construed broadly, will tend to encompass all claims which are technically eligible for fund financing. If construed in a narrow sense, then only contamination sites that actually receive funds will be eligible, even though the EPA has not indicated a valid interpretation as to its intention in interpreting the phrase. In *Exxon* v. *Hunt*,[79] the plaintiffs, a consolidated group of chemical and oil companies, challenged the validity of the New Jersey Spill Fund. Plaintiffs, hereinafter referred to as Exxon, sought an injunction against further collection of the tax which finances the Spill Fund, and a refund of all sums paid into the fund. The court, for several reasons, rejected Exxon's argument that the fund is entirely preempted by the operation of Superfund's Subsection 114(c).

At the time of this writing, the U.S. Supreme Court heard arguments on this case on December 9, 1985. No decision has yet been rendered. The New Jersey Supreme Court in June of 1984 upheld the collection of funds under the New Jersey Spill Compensation Fund. At the present time, the fund can be used to pay not only for clean-up costs and claims not actually paid under CERCLA, but for New Jersey's share of federally financed clean-ups as well. The New Jersey Supreme Court, in upholding the tax first stated that the legislative history of Superfund suggests that Congress intended to construe Subsection 114(c) narrowly. Although Superfund was passed in a last minute compromise, and not subject to a conference committee report, the phrase was sufficiently discussed on the House and Senate floors.

Secondly, the court found that the New Jersey Spill Fund was much more encompassing than Superfund's fund.[81] The New Jersey Spill Fund is strictly liable for direct and indirect damages without regard as to who sustained the damages. Also, the New Jersey Spill fund provides for compensation of victims with regard to property damage and lost income excluding at date medical expenses.

Lastly, the court said Congress did not view Superfund as the ultimate solution to the problems of hazardous waste clean-up. The total cost of cleanig up the sites on a national level will be an estimate between $10 and $20 billion. A fund of $1.6 billion is insufficient to handle an insurmountable problem such as this. States must play a significant and ongoing role in the clean-up operation.

In the final analysis, the preemption clause found in Subsection 114(c) tends to restrict the remedies available for financing the clean-up of hazardous waste sites. The imposition of fees and taxes can result in the taxpayers absorbing the cost of clean-ups. Congress and the New Jersey legislation have stated that the cost should fall upon the industry which has for years benefited from uncontrolled, inexpensive, and unsound environmental waste disposal techniques.

Finally, preemption also may occur at the state-local interfacing of legislation. Preemptive capabilities and the limitations of this secondary level are easy to overlook. Much of the doctrinal analysis outlined above is directly applicable to the invalidation of inconsistent local law.

A North Carolina District Court in *U.S.* v. *Waste Industries*[83] specifically reached this issue of preemption, holding that the common law of nuisance is preempted by RCRA because of the comprehensiveness of the regulatory program and this area being supervised by an expert adminstrative agency. The *Waste Industries* case included a Section 7003 action seeking an abandoned landfill. In holding, Section 7003 of RCRA does not apply to an abandoned landfill suspected of causing ground-water contamination.

The multitude of various factors indigenous to a locality, coupled with the wide array of state regulatory schemes, creates a patchwork of preemptive possibilities. Therefore, it is difficult to be more specific on this issue without closely examining the particular legal components involved in preemption. The existence of home rule in many states, for example, is a significant factor affecting local preemptive capability.

The continued encouragement of local experimentation in the alleviation of environmental problems has been discussed. This attitude establishes a presumption of the validity of such local remedial legislation. This effect is underscored and magnified by the fact that state courts construe and interpret these local laws; these courts, whenever possible, avoid finding irreconcilable conflict among the respective legislation. Preemption is not limited to the state-federal level, although it is often thought to occur only at this level.

In another preemption case, the New Hampshire Supreme Court struck down a local ordinance which allowed the issue of whether a hazardous waste facility may be constructed to be decided by popular town vote. The Court in *Stablex Corp.* v. *Town of Hooksett*[84] invalidated this ordinance because it conflicted with New Hampshire state law. The Court expressly rejected arguments by the municipality that this ordinance allowing for popular vote was within the authority and preview of "home rule", stating "[the New Hampshire State Plan was] intended to be implemental on a state-wide basis. As such, it completely preempts the field on hazardous waste in this state."

The final section explores the role of the commerce clause in restraining abuses in state legislation.

V. THE COMMERCE CLAUSE

A. Introduction

The modern analysis of the commerce clause uses a balancing of interests test to determine whether particular state legislation unduly burdens interstate commerce.

The functional importance of a court's impartiality in balancing these competing interests allows it effectively to curb abuses of individual police power. "The judiciary plays an important role in assessing the validity of interstate commerce regulation by the states, because it may be the branch of government most able to provide a check against excessive local assumption of power. When a state legislature determines how much power it has to affect interstate commerce, there may be no inner political check on its action.[85]

For example, the freedom of a state to enter into an agreement with another state is constitutionally circumscribed by the compact clause.[86]

Historically, certain sections of the constitution have received greater judicial attention than others. The compact clause, for instance, has lacked both consistency and effectiveness in its interpretation as an expressed prohibition of state agreements.[87] Further, case law in the last 50 years has given the commerce clause[88] an expansive interpretation that has relegated the compact clause to a secondary status. The preeminence of the commerce clause in terms of constitutional doctrine is no less apparent within the area of solid waste.[89]

The authority of the commerce clause primarily has been used to justify federal environmental legislation.[90] "Whatever amounts to more or less constant practice and threatens to obstruct or unduly burden the freedom of interstate commerce is within the regulatory power of Congress under the commerce clause, and it is primarily for Congress to decide.[91]

Furthermore, the notable absence of a substantive analysis of the preemption issue in *City of Philadelphia,* coupled with the Court's application of commerce clause principles, is understandably a strong precedent that suggests that hazardous waste legislation will be influenced strongly by commerce clause analysis. In fact, several courts have held state laws to be a violation of the commerce clause which attempted to exclude out of state waste.

The commerce clause has not only a positive effect, i.e., authorizing Congress to enact laws regulating interstate commerce, but also a negative impact in forbidding, on its own, without congressional action, certain kinds of state interference with interstate commerce. This is the "negative" or "dormant" power of the commerce clause. This dual impact was stated clearly in *Freeman* v. *Hewit.*[92]

Freeman echoes the classic case of *Cooley* v. *Board of Wardens* which prohibits states from regulating subjects that are inherently national in scope.[93]

Historically, the threshold issue decided was whether the subject (or article) was an article of commerce, which often determined whether a state could legislatively ban or strictly regulate it. On the judicial finding that the article merited commerce clause protection, a balancing of the local benefit derived from the state legislation in view of its concomitant burden on interstate commerce was required. The Supreme Court's language in *City of Philadelphia* clearly has repudiated this simplistic listing approach of commerce clause exclusion, as noted below.

Although this approach was rejected, a long line of authority holds that a state reserves the right to enact "quarantine legislation" prohibiting the importation or transportation of objects that spread disease or are inherently harmful. The basis for such state authority is its reserved powers, mentioned above. This is found in Article I, Section 10 of the Constitution and crystallized via the Tenth Amendment.

The leading case supporting the state's right to enact quarantine legislation is *Sligh* v. *Kirkwood.*[94] Arguments were advanced that local interests prevent articles that are inherently injurious to public health, safety, and welfare from being transported into a state. The following language, taken from *Sligh,* is instructive of this exclusionary approach to the commerce clause:

> The power of the state to prescribe regulations shall prevent the production within its borders of impure foods, unfit for use, and such articles as would spread disease and pestilence, is well established. Such articles, it has been declared by this court, are not the legitimate subject of trade or commerce, nor within the protection of the commerce clause of the Constitution...Nor does it make any difference that such regulations incidentally affect interstate commerce when the object is to protect the people of the state.[95]

Similarly, alfalfa infected with boll weevil,[96] the transportation of diseased cattle,[97] and the banning of impure foods unfit for consumption also were determined sufficiently harmful to warrant prohibition under the reserved police power of the states. In *Clason* v. *Indiana,*[98] the mandatory disposal of large dead animals within 24 hr of death was upheld on the basis that such carcasses were not legitimate objects of commerce. In *Clason,* the Supreme Court affirmed the Indiana Supreme Court's finding that "dead bodies of animals slaughtered for food are not legitimate subjects of Commerce." The Court's finding that "the obvious purpose of the enactment (was) to prevent the spread of disease and the development of nuisances" was significant in upholding the state's exclusion of such objects as legitimate articles of commerce.

The analysis of the majority in *City of Philadelphia* casts doubt over the vitality of this exclusionary approach to the commerce clause. Justice Stewart, speaking for a 7—2 majority in this 1979 case, stated: "All objects of interstate trade merit commerce clause protection; none is excluded by definition at the outset." The adoption of this reasoning, therefore, directs inquiry into the burden it creates on interstate commerce in view of the local benefit derived.

The result reached in *City of Philadelphia* through the use of balancing principals suggest that, in the final analysis, this new reasoning may have little effect; in *City of Philadelphia* the Court rejected the appellee's argument that the statutory provision was a valid quarantine measure.[99] The dissenting opinion, filed by Justice Rehnquist and joined by Justice Burger, advocated that the New Jersey law was, in fact, a valid quarantine measured comparable to other prohibitions sustained against noxious items, characteristic of *Clason* and *Sligh*.[100]

B. Balancing and the Principle of Nondiscrimination

Despite the above dissent, the substance of the majority opinion focuses on a balancing of the conceded impact on interstate commerce with the respective interest of the state in regulating or prohibiting the matter at hand. "[W]here other legislative objectives are credibly advanced and there is no patent discrimination against interstate trade, the Court has adopted a much more flexible approach, the general contours of which were outlined in *Pike v. Bruce Church, Inc.*"[101] The balancing test, as enunciated in *Pike,* states:

> Where the statute regulates evenhandedly to effectuate a legitimate local public interest, and its effect on interstate commerce are only incidental, it wil be upheld unless the burden imposed on such commerce is clearly excessive in relation to the putative local benefits...If a legitimate local purpose is found, then the question becomes one of degree.[102]

An examination of the above language in *Pike* suggests that state regulatory legislation must satisfy four separate criteria to survive commerce clause scrutiny. The legislation must: (1) not discriminate against interstate commerce; (2) further a legitimate local public interest; (3) have only an incidental effect on interstate commerce; and (4) not be excessively burdensome compared to the local benefit received.

In *Pike,* Arizona required that all cantaloupes grown in the state be packaged in standard containers identifying the fruit as being Arizona. The Court held that this labeling requirement was overly burdensome in view of the state's purpose of enhancing the reputation of the quality of Arizona grown cantaloupes. Thus, in so finding an "undue burden on commerce", *Pike* does not provide specific guidelines probative of what such a "legitimate local purpose" might be; such an analysis is intrinsically fact-specific and will be judicially determined on a case-by-case basis.

The balancing test of *Pike* and the safeguards against overly burdening interstate commerce are equally applicable to ordinances; that is, the ordinance must address a "legitimate local public interest". It is difficult to speculate what contours exist in this limitation, although it is clear that any statute or political subdivision contravenes this mandate and risks invalidation. This is the rule as set forth in *Huron Portland Cement Co.* v. *Detroit,*[103] where a city ordinance regulating smokestack emissions was held to be a valid exercise of state police power and not a violation of the commerce clause. In upholding Detroit's regulation of smoke emissions from vessels licensed to operate in interstate commerce, the Court noted the close nexus between the purpose of environmental protection and the traditional police power objectives of health and welfare. The majority found that "the sole aim of the Detroit ordinance was the elimination of air pollution to protect the health and enhance the cleanliness of the local community."[104] In noting that the ordinance had no discriminatory effect, the Court concluded that the city or state environmental regulation directed at enhancing air quality "clearly falls within the exercise of even the most traditional concept of what is compendiously known as the police power."[105] Thus, the Court (Justice Stewart) sustained this commerce clause challenge, despite its conceded impact on interstate commerce. Environmental ordinances are accorded a "preemptive validity" in attempting to rectify what is often a local environmental problem, although the Court in *Huron* declined to speculate on the cumulative burden and effects on other cities if they enacted similar ordinances.

Any ordinance or regulation that yields an economic advantage on the state or political subdivision is clearly outside the contour of a "legitimate local public interest". Courts are sensitive to such an improper impact and will carefully examine the facts or record for probable economic protectionist effects.The Supreme Court has repeatedly invalidated state regulations on commerce clause principals. Typically, those laws that run afoul of this nondiscrimination mandate attempt to strictly regulate (or have in some cases, even prohibit importation of) out-of-state milk under the guise that the enactment was meant to protect the health of its citizens.[106]

This mandate against a state enacting an economic protectionist law contrary to the commerce clause was recently applied in the *City of Philadelphia* v. *New Jersey,* mentioned earlier.[107] In *City of Philadelphia,* the appellants strenuously maintained that New Jersey Law P.L. ch. 363 was motivated by financial considerations and hence was an improper protectionist measure, the Supreme Court agreed, holding the New Jersey law unconstitutional as a violation of the commerce clause.

As with preemption, a court will examine the statute's stated purpose, as well as its method of accomplishing this purpose to determine whether legislation is "basically a protectionist measure, or whether it can be viewed as a law directed to legitimate local concerns, with effects upon interstate commerce that are only incidental."[108] A finding that the local interest could be regulated with less impact on interstate commerce by an alternative means is a sufficient basis to hold the legislation violative of the commerce clause. Thus, the burden is on the state to demonstrate that other alternatives have been considered. The court noted this burden of proof in *Hunt* v. *Washington Apple Advertising Commission:*

> When discrimination against commerce of the type we found here is demonstrated, the burden falls on the state to justify it both in terms of the local benefits flowing from the statute and the unavailability of nondiscriminatory alternatives, adequate to preserve the local interest at state.[109]

In *City of Philadelphia,* the legislative means chosen (i.e., discriminating against out-of-state waste) was too burdensome in view of the statute's otherwise legitimate purpose of reducing the rate of waste flowing into New Jersey landfills. In other words, a state may not prohibit out-of-state waste unless there is some reason, apart from the waste's origin, for the prohibition. The following language from *City of Philadelphia* summarizes this principle of nondiscrimination:

> It may be assumed as well that New Jersey may pursue those ends by slowing the flow of all waste into the State's remaining landfills, even though interstate commerce may incidentally be affected. But whatever New Jersey's ultimate purpose, it may not be accomplished by discriminating against articles of commerce coming from outside the state unless there is some reason, apart from their origin to treat them differently. Both on its fact and in its plain effect, ch. 363 violates this principle of nondiscrimination.[110]

As noted in the above case, New Jersey P.L. 1973, ch. 363 contained a flat prohibition against the importation of most out-of-state waste. Several cases subsequent to the *City of Philadelphia* case have clarified this principal of nondiscrimination. The Tenth Circuit Court of Appeals extended the holding enunciated in *City of Philadelphia* in *Hardage* v. *Atkins.*[111] The court in *Hardage* found the challenged Oklahoma statute violated the commerce clause where it required that a sister state maintain "substantially similar standards" before out-of-state waste could be disposed in Oklahoma.[112] The court rejected appellant's contention that this "substantially similar" criterion was less restrictive than the invalidated provision found in the New Jersey statute.[113]

The lower court, before being reversed by the Tenth Circuit, adopted the position that industrial waste was not an item of commerce, thus upholding Oklahoma's statute. Upon appeal, the Court of Appeals found the Oklahoma reciprocity provision unconstitutional, implying that no rational relationship existed between state public health and safety. Upon appeal after remand, the Court of Appeals found that Oklahoma's provision requiring "substantially similar standards" amounted to discriminatory economic protection and was therefore invalid and unconstitutional.

Before closing, it is appropriate to briefly mention that an alternative judicial analysis has been occasionally used in deciding whether state laws pass constitutional muster under the commerce clause. This alternative to this balancing test characteristic of *Pike* has been utilized where a court has found it particularly difficult to quantify commercial interests against such noncommercial (and inherently nonquantifiable) interests as safety.

The traditional balancing test of *Pike* has two major shortcomings. First, it presumes that state interests and competing National Interstate Commerce interests are quantifiable and hence have the inherent ability to be balanced. The second criticism of this test is related to the first. To effectuate a meaningful balance, some sort of common scale of measurement is required. State environmental interests, more often than not, arise out of unique circumstances, making it unlikely that any meaningful standard will ever develop in this area. In such cases, a court's attempt to "balance" state against national interests is reduced to a subjective value judgment resulting from the particular persuasion of the court's membership at that time.

The scrutiny applied is less intense in this alternative test. It is limited to a twofold judicial examination. First, the statute or ordinance must address a legitimate state purpose. This will include an examination of resulting discriminatory effects, if any. Second, the court will examine the means selected by the legislature in its approach to addressing the problem. This second aspect is essentially limited to a determination whether the means selected is "rationally related" to the state purpose. One concept developed at length in this discussion chapter is that courts will give legislatures wide discretion in selecting the means chosen to abate environmental problems.

The above test was first affirmatively applied in *Brotherhood of Locomotive Fireman and Engineers* v. *Chicago, Rock Island & Pacific Railroad*,[116] where the court refused to apply a balancing test and recognized the incongruity of measuring safety interests against commerce interests. Significantly, the court in *Brotherhood* was critical of the lower court for attempting to balance "financial losses...against the loss of lives and limbs of workers and people using the highways."[117]

Parties advocating a court's use of this alternative test often argue that Congress (or state legislatures) are better able to balance such incongruent interests, since the latter are not bound to the particular facts before them.

This alternative approach to strict balancing has received some judicial support in its application to judicial reviewability of state environmental laws under the commerce clause. The vitality of this approach is best phrased in the negative; at least two cases since *Brotherhood* have *refused* to apply a balancing test. In *American Can Co.*, discussed earlier, the Court commented upon the "inappropriateness of a weighing process in cases of noncomparable benefit and injury."[118] Similary, in *Soap and Detergent Association* v. *Clark,* a balancing test was rejected.[119] While the observation that in all three of the above cases the state statutes were upheld as valid may or may not be significant, it is clear that less judicial scrutiny accompanies this alternative test. In justifying this lower judicial posture in applying this test, it is often considered that Congress is better able to resolve the scope of state environmental regulation than the judiciary.

The shortcomings aside, the continual growth of this alternative test as a viable means of passing upon state environmental legislation remains uncertain. The court in *City of Philadelphia* declined the reasoning of *Brotherhood* and instead applied a strict balancing test in invalidating the New Jersey law.

VI. CONCLUSION

This chapter has outlined the legal framework in which both RCRA and Superfund have been enacted and will bear scrutiny. An appreciation of the substantive nature and source of this constitutional authority assists in placing this regulatory legislation in perspective.

Their impact will continue to generate widespread interest. The judicial scrutiny applied within this legal framework has been examined in some detail, both in terms of traditional due process and equal protection criteria. The importance of these concepts has warranted discussion, despite the observation that an argument probably will not be meritorious on either of these constitutional grounds. This observation is buttressed by both the prevailing "minimum rationality" standard and the court's long-standing deference to the legislature in remedying economic displacements, as illustrated by *Lee Optical* and more recently in *American Can* v. *Oregon Liquor Control Commission.*

Through clear Congressional directive, for example, the small generator exemption may be rationally lowered from 1000 to 100 kg/month.

No less important in providing a statutory standard for judicial review of federal agency regulations is the Administrative Procedure Act. Discussion of this issue acknowledged the significant impact of the *Vermont Yankee* decision in 1978, as typified by the posture of the D.C. Court of Appeals in its *Weyerhauser* decision.

Much emphasis has been placed on preemption, which is of primary importance in the assessment of the validity and pervasiveness of state legislation in the federal context, and local legislation in the state context. It is unusual for legislation specifically to state the preemptive parameters for which it was enacted. Where an expressed purpose is indicated, the subordinate legislation will be displaced. This rarely occurs. Instead, it is the courts that must determine the extent to which the two Acts are compatible; such an inquiry includes a thorough examination of the Act's legislative history to glean legislative intent.

The summary dismissal of this issue in *City of Philadelphia* must be criticized, both because of its historical and doctrinal importance, and because it is inconsistent with past preemption decisions.

The 1984 amendments create a series of unanswered questions as to state authorization plans. Section 228 appears to be a restriction from the 1980 amendment to Section 3009 which authorized more stringent local regulatory standards than what was minimally required under RCRA. Clearly, this 1980 amendment created the potential for widespread state-state and state-federal regulatory conflict in terms of the overall good of a regional approach to hazardous waste management.

Finally, the broad authority of the commerce clause has been the constitutional mainstay most often used to justify federal environmental legislation. Its modern application involves a "balancing of interests" test, as illustrated in *Pike.* Presently, the commerce clause is the cutting edge in restraining abuses of state power as reflected in *City of Philadelphia* and *Hartage.* The Court's vital role as an impartial watchdog in invalidating local laws that discriminate against out-of-state interests has been noted. The observation that Oklahoma's substantially similar standards provision was invalidated suggests that state governments may face difficulty in their localized efforts to regulate this subject area.

Obviously, this subject matter is exceedingly complex. The commerce clause will be on the constitutional forefront restraining the conflict, discrimination, and abuses arising from these complexities, as these two Acts and regulations issued thereunder embark on a previously unattained level of regulatory compliance.

FOOTNOTES

1. "The Committee believes that the approach taken by this legislation eliminates the last remaining loophole in environmental law, that of unregulated land disposal of discarded materials and hazardous wastes. Further, the committee believes that this legislation is necessary if other environmental laws are to be cost effective." H.R. 94-1491, p. 4. Reprinted in (1976) U.S. Code Cong. & Ad. News 6241.

2. Sec. 228, H.R. 2867.

3. See 130 *Cong. Rec.* Hll, 103-44 (daily ed., Oct. 3, 1984) and 130 *Cong. Rec.* S13, 812-23 (daily ed., Oct. 5, 1984); H.R. 2867, 98th Cong., 2d Sess., 130 *Cong. Rec.* Hll, 103-25 (daily ed., Oct. 3, 1984) (hereinafter "H.R. 2867").

4. Sec. 221, H.R. 2867.

5. For example, A finding made by Congress states: "[C]ertain classes of land disposal facilities are not capable of assuring long-term containment of certain hazardous wastes, and to avoid substantial risk to human health and the environment, reliance on land disposal should be minimized or eliminated, and land disposal, particularly landfill and surface impoundment, should be the least favored method for managing hazardous wastes." These amendments also contain limitations as to land disposal of particular hazardous wastes. (Sec. 201, H.R. 2867).

6. Sec. 213, H.R. 2867.

7. Sec. 204, H.R. 2867.

8. See 40 C.F.R. Sec. 261.6 (1983).

9. Sec. 204, H.R. 2867.

10. *Id.* This provision is to take effect 90 days after signing of the Act and is self implementing.

11. Sec. 401, H.R. 2867.

12. Sec. 601, H.R. 2867.

13. This program is particularly ambitious since all "Superfund" (e.g., The Comprehensive Environmental Response, Compensation and Liability Act of 1980, as amended, 42 U.S.G. Section 961 *et. seq.*) regulated substances are subject to this Amendment. (A "regulated substance" is any *Superfund* hazardous substance (except hazardous waste) and petroleum.)

14. E.g., see Sec. 201.

15. These provisions allow EPA time in which to promulgate regulations. However, if EPA fails to issue the regulation, the regulated community must comply with the regulation enacted by Congress.

16. "Superfund", The Comprehensive Environmental Response, Compensation And Liability Act of 1980, as amended, 42 U.S.C. Sec. 9601, *et seq.*

17. Sec. 206, H.R. 2867 will extend RCRA to include subsurface contamination predating the effective date of EPA's corrective action requirements. See 40 C.R.F. Sec. 264.90 and 264.98-264.100 (1983).

18. *Shell Oil Co.* v. *EPA* (D.C. Cir. No. 80-1532).

19. See 45 *Fed. Reg.* 55, 386 (Aug. 19, 1980); 45 *Fed. Reg.* 74, 489 (Nov. 10, 1980); 46 *Fed. Reg.* 60, 446 (Dec. 10, 1980); and 47 *Fed. Reg.* 15, 307 (April 8, 1982).

20. 90 Stat. 2809. Public Law 94-580 (passed October 21, 1976), as amended by the Quiet Communities Act of 1978. 42 U.S.C. 6901 *et. seq.*, as amended by the Solid Waste Disposal Act Amendments of 1980. Public Law 96-482 (passed October 21, 1980; as amended by the Hazardous and Solid Waste Amendments of 1984 (H. Rep. No. 98-1133). The Solid Waste Disposal Act of 1965, 79 Stat. 998 (1965), was the first federal legislation that directly addressed the issue of solid waste. The thrust of the Act was research into and development of improved methodologies of solid waste processing.

21. 45 *Fed. Reg.* 12746 (Feb. 26, 1980). This observation by EPA also is echoed in the Act's legislative history: "Hazardous wastes typically have little, if any, economic value; are often not susceptible to neutralization; present serious danger to human life and environment; and can only safely stored, treated and disposed of at considerable cost to the generator." H.R. 94-1491, p. 4. Reprinted in (1976) U.S. Code Cong. & Ad. News 6241.

22. "It should be noted that discarded materials are generated from a multitude of sources in every sector of the nation's life. The committee recognizes among those sources the pollution abatement activity initiated as a result of federal air and pollution laws. In summary, discarded materials are a direct result of national industrial production and the American life style." H.R. 94-1491 part one, p. 3. Reprinted in (1976) U.S. Code Cong. & Ad. News 6240.

23. As one commentator has explained in this regard: "There is a basic difference between land on the one hand and air and water on the other which accounts for their different positions in environmental problems. It is technically relatively easy to define property rights in land...It is difficult to establish property rights in air and water and they are part of what is usually referred to as the public domain." Mills, *The Economics of Environmental Quality,* W.W. Norton & Co., New York, 1978, 33.

24. Sec. 101, H.R. 2867.

25. *Cong. Rec.* 10/5/1984 (remarks of Sen. Chafee), S. 13818.

26. Sec. 201, H.R. 2867; EPA is required to accomplish this review of wastes over a $5^1/_2$-year period.

27. Sec. 202, H.R. 2867.

[28] RCRA §1004(5), 42 U.S.C. §6903(5)(1976).

[29] 40 C.F.R. Sec. 261.5 (1983).

[30] This list is located at 40 C.F.R. part 261.31.

[31] Sec. 233, H.R. 2867. Within 270 days of the date of passage, generators of "greater than 100 but less than 1000 kilograms of hazardous wastes per month must provide for proper identification of such wastes which are transported off-site. Such generators...have to notify transporters that such wastes are hazardous but would not have to comply with full hazardous waste testing. *Cong. Rec.* 10/5/1984 (remarks of Sen Chafee), p. 13818.

[32] 40 C.F.R. part 262.

[33] 40 C.F.R. part 122.

[34] *NRDC* v. *EPA* (D.C. Cir. No. 80-1607).

[35] Safe Drinking Water Act, 42 U.S.C. §300f *et. seq.*

[36] H.R. Rep. No. 94-1491 (Interstate and Foreign Commerce Committee).

[37] S.R. 94-869 and S.R. 94-988 (Public Works Committee).

[38] *Fed. Reg.* Vol. 43, No. 243 p. 58948 (Dec. 12, 1978).

[39] Conference Report (H. Rep. No. 98-1433).

[40] Section 1003 outlines the Act's objectives. "The objectives of this Act are to promote the protection of health and the environment and to conserve valuable material and energy resources by (1) providing technical and financial assistance to State and local governments and interstate agencies for the development of solid waste management plans (including resource recovery and resource conservation systems) which will promote improved solid waste management techniques (including more effective organizational arrangements), new and improved methods of collections, separation, and recovery of solid waste, and the environmentally safe disposal of nonrecoverable residues...(8) establishing a cooperative effort among the Federal, State, and local governments and private enterprise in order to recover valuable materials and energy from solid waste."

[41] 42 U.S.C. §6926.

[42] Sec. 228.

[43] The Comprehensive Environmental Response, Compensation and Liability Act of 1980, As Amended, 42 U.S.C. §9601 *et. seq.* Also see H.R. Rep. No. 1016, 96th Cong. 2d. Sess. 1 reprinted in 2 ELI, Superfund: A Legislative History 429 (1982).

[44] *U.S.* v. *Northeastern Pharmaceutical and Chemical Co., Inc.,* No. 80-5006 (19ERC 2186) (decided September 30, 1983).

[45] H.R. 5640 (1984).

[46] Through the years, the Court has expanded the definition of commerce. See *Heart of Atlanta Motel, Inc.* v. *United States,* 379 U.S. 241 (1964); *Wichard* v. *Filburn,* 317 U.S. 111 (1942). "The meaning of the commerce clause for state power has been extensively debated and revised over time. Historians have attributed the existence of the clause to the profound concern over the erection of trade barriers between the states which had inhibited interstate trade under the Articles of Confederation." See, generally, Hutchinson, *The Foundations of the Constitution,* University Books, Inc., chap. 8 (1975); Schwartz, *A Commentary on the Constitution of the United States,* Rothman & Co., chap. 6 (1963); Antieau, *Modern Constitutional Law,* Vol. 2, Bancroft Whitney Co., San Francisco, 1969, 4.

[47] *Id.*

[48] "Lacking other constitutional language that was as conveniently flexible, the courts have turned to the commerce clause to sustain all manner of federal regulation, and environmental regulation is no exception. The states retain their traditional powers over commerce, health, and safety but, backed by the courts. Congress has steadily enlarged the federal role in environmental protection through increasing reliance upon the commerce power." Anderson *Environmental Improvement Through Economic Incentives,* Johns Hopkins Univ. Press, Baltimore, 1977, 108.

[49] U.S. Const. Art. I., Sec. 8., cl. 18. The so-called implied power of Congress authorizes Congress "To make all laws which shall be necessary and proper for carrying into Execution the foregoing Powers, and all other Powers vested by this Constitution in the Government of the United States, or in any Department or Officer thereof."

[50] "For the National Government, federalism is rarely an obstacle to action. Relying on its power to regulate interstate commerce, Congress can enact legislation that it determines will further the general welfare. National legislators may reasonably conclude that the public benefit of compulsory antipollution measures exceeds the sum of the cost to each citizen of compliance with such laws. Therefore, unless there is no rational basis for the congressional determination that the pollution sought to be regulated affects interstate commerce, the balance struck by Congress in enacting antipollution legislation is very likely to withstand judicial scrutiny." Comment, "State Environmental Protection Legislation and the Commerce Clause" 87 Harv. L. Rev. 1762, 1771 (1974).

[51a] No. s-82 — 1913, Vol. 82-1951, 53 Law Week 4135 (Feb. 19, 1985).

[51b]   The Garcia decision overruled *National League of Cities v. Usery* 426 U.S. 833 (1976), where the court ruled that the Tenth Amendment prohibited Congress from using its Commerce Power "to force directly upon the states its choices as to how essential decisions regarding the conduct of integral governmental functions are to be made" (at 855).

[52]   *Duke Power Co. v. Carolina Environmental Study*, 483 U.S. 59, 83 (1978).

[53]   *Williamson v. Lee Optical of Oklahoma*, 348 U.S. 483 (1955).

[54]   Id., at 487-488. See also *Washington Star Co. v. International Typographical Union Negotiated Pension Plan*, 729 F. 2d 1502 (C.A.D.C. 1984) (holding that it is for the legislative and not the courts to evaluate the rationality of economic legislation).

[55]   *American Can Co. v. Oregon Liquor Control Commission*, 517 P. 2d. 691(1973). Also see *National Treasury Employees Union v. Devine*, 591 F. Suppl. 1143 (D.C.D.C. 1984) (legislative acts adjusting burdens and benefits of economic life have a presumption of constitutionality).

[56]   See footnote 50 above at 1768.

[57]   National Environmental Policy Act 42 U.S.C. 4321-4361 (1970).

[58]   Contrast: *Calvert Cliffs Coordinating Committee v. Atomic Energy Commission*, 449 F. 2d. 1109 (D.C. Cir. 1971), *Environmental Defense Fund v. Corps. of Engineers*, 470 F. 2d. 289 (8th Cir. 1972), and *NRDC v. Morton*, 458 F. 2d. 827 (D.C. Cir. 1972) with *Vermont Yankee Nuclear Power Corp. v. NRDC* 435 U.S. 519 (1978) and *Strycker's Bay Neighborhood Council, Inc. v. Karlen*, 444 U.S. 1307 (1980).

[59]   Administrative Procedure Act (5 U.S.C. §701-706).

[60]   Section 706 of the Administrative Procedure Act provides that the reviewing Court shall: ...(2) Hold unlawful and set aside agency action, findings, and conclusions found to be (A) arbitrary, capricious, an abuse of discretion, or otherwise not in accordance with law; (B) contrary to constitutional right, power, privilege, or immunity; (C) in excess of statutory jurisdiction, authority, or limitation or short of statutory right; (D) without observance of procedure required by law...; or (F) unwarranted by the facts to the extent that the facts are subject to trial de novo by the reviewing Court. (5. U.S.C. Section 706.)

[61]   *Vermont Yankee Nuclear Power Corp. v. National Resources Defense Council*, 435 U.S. 519, 555 (1978).

[62]   *Vermont Yankee* at 557.

[63]   *Weyerhaeuser v. Castle*, 590 F. 2d. 1011, 1028 (D.C. Cir. 1978).

[64]   14 ERC 1189.

[65]   At 1200.

[66]   "[i]t is settled...that the state may not...under the guise of exercising its police power or otherwise, enact legislation in conflict with the statutes of Congress passed for the regulation of the subject, and, if it does, to extent that the state law interferes with or frustrates the operations of the acts of Congress, its provisions must yield to the superior Federal power given to Congress by the Constitution." *McDermott v. Wisconsin*, 288 U.S. 115 (1913).

[67]   Comment, "Preemption as a Preferential Ground: A New Canon of Constitution," 12 Stan. L. Rev. 208, 215 (1959).

[68]   437 U.S. 617 (1978).

[69]   The statutory provision in question was N.J.P.L. 1973 ch. 363, which provided: "No person shall bring into this State any solid or liquid waste which originated or was collected outside the territorial limits of the State, except garbage to be fed to swine in the State of New Jersey, until the commissioner (of the State Department of Environmental Protection) shall determine that such action can be permitted without endangering the public health, safety and welfare and has promulated regulations permitting and regulating the treatment and disposal of such waste." N. J. Rev. Stat. Sec. 13:11-10. In February 1974, the commissioner released regulations as authorized by ch. 363, permitting the following four categories of waste to enter New Jersey: (1) garbage to be fed to swine: (2) "Pesticides, hazardous waste, chemical waste, bulk liquid, bulk semi-liquid, which is to be treated, processed or recovered in a solid waste disposal facility which is registered with the Department for such treatment, processing or recovery, other than by disposal on or in the lands of this State." N.J.A.C. 7:1-4.2

[70]   See, generally: Comment, "The Preemption Doctrine: Shifting Perspectives on Federalism and the Burger Court," 75 Colum L. Rev. 623, 653 (1975). See also Danaher, "Waste Embargo Held a Violation of the Commerce Clause," Conn. L. Rev., Vol. 11, No. 2, 292, 306 (winter, 1979).

[71]   *Jones vs. Rath.* 430 U.S. 519, 526 (1977).

[72]   At 52.

[73]   This issue of state sovereignty has resulted in a severely divided court. See *Garcia v. Samta, supra*; *National League of Cities v. Usery*; Margulies, Discord on the Court, 10 Conn. L. Trib. No. 22 (May 28, 1984).

[74] In *Jones,* a California statute that concurrently regulated the method of labeling of flour (Section 12211 of the Cal. Bus. and Prof. Code: West Supp. 1977) was successfully challenged on preemption grounds. Although the decision "hold(s) that 15 U.S.C. Section 1461 (Fair Packaging and Labeling Act) does not preempt California's Section 12211 "the preemption challenge was sustained in that enforcement of section 1211 "...would prevent 'the accomplishment and execution of the full purposes and objectives of Congress' in passing the FPLA." At 540, 544.

[75] At 526.

[76] H.R. 94-1491, p. 33. Reprinted in (1976) U.S. Code Cong. & Ad. News 6271.

[77] Section 3009 provides: "[U]pon the effective date of regulations under this subtitle no State or Political Subdivision may impose any requirements less stringent than those authorized under this subtitle..."

[78] The Act also authorizes the administrator, "after notice and opportunity for public hearing, [to] withdraw his approval of such plan. Such withdrawal of approval shall cease to be effective upon the Administrator's determination that such complies with such minimum requirements." Section 4007.

[79] *Exxon* v. *Hunt,* No. SC 303A-81 (N.J. Tax Ct. filed Aug. 12, 1981). A number of related cases have also been filed. In *Exxon* v. *Hunt.* No. 81-1458 (D.N.J. Aug. 10, 1981) (order of dismissal), *appeal docketed,* No. 81-2514 (3rd Cir. Aug. 12, 1981), the defendants successfully argued that the federal Tax Injunction Act barred consideration by the federal district court of whether CERCLA preempts the New Jersey Act.

[80] Senator Jennings of Virginia, in CERCLA's legislative history approved of the statement that "state funds are preempted only for efforts which are in fact paid for by the federal fund that there would be no preemption for efforts which are eligible but for which there is no reimbursement" 126 Cong. Rec 514, 981 (daily ed. Nov. 24, 1980).

[81] Compare N.J. Stat. Ann. §58; 10-23.11g(a) (West 1982) with 42 U.S.C.A. §9601(14) (West 1980) Laws Special Pamphlet.

[82] 775 F.2d 627 (1985).

[83] No. 80-04, decided Jan. 3, 1983 N.C. (1983).

[84] No. 82-220 N.H. (1984).

[85] See Note 50, *supra* at 1777. Also see Tushnet, *Rethinking the Dormant Commerce Clause,* 55 Wis. L.R. 125 (1979).

[86] The Compact Clause (Art. I. Sec. 10, cl. 3) states, "No State shall, without the consent of Congress...enter into any Agreement or Compact with another State or with a foreign power..." This prohibition has been addressed specifically in section 1005(b) of RCRA, in which Congress has authorized interstate agreements to facilitate the goals of this legislation. U.S.C. 6904.

[87] See, generally: Engdahl, "Characterization of Interstate Arrangements," 64 Mich. L. Rev. 63, 67-73 (1965); for a historical discussion see Weinfeld, "What Did the Framers of the Federal Constitution Mean by 'Agreements or Compacts'?" 3 U. Chi. L. Rev. 453 (1936).

[88] Art. I. Sec. 8, cl. 3.

[89] In his book, *Environmental Improvement Through Economic Incentives* (Johns Hopkins University Press, Baltimore, 1977, Anderson remarks on the steadily increasing role the commerce clause has played in the justification of environmental legislation: "Lacking other constitutional language that was as conveniently flexible, the courts have turned to the commerce clause to sustain all manner of federal regulation, and environmental regulation is no exception..., backed by the courts, Congress has steadily enlarged the federal role in the environmental protection through increasing reliance upon the commerce power" (at 10).

[90] See Comment, footnote 50 at 1764.

[91] *Stafford* v. *Wallace,* 258, U.S. 495 (1922).

[92] In *Freeman* v. *Hewit* 329 U.S. 249 (1946) the Court stated: "...[T]he Commerce Clause was not merely an authorization to Congress to enact laws for the protection and encouragement of commerce among the States, but by its own force created an area of trade free from interference, by the States. In short, the Commerce Clause even without implementing legislation by Congress is a limitation upon the power of the States. (Citations omitted)...A State is also precluded from taking any action which may fairly be deemed to have the effect of impeding the free flow of trade between States..." (at 252).

[93] *Cooley* v. *Board of Wardens,* 53 U.S. 299 (1851).

[94] *Sligh* v. *Kirkwood,* 237 U.S. 52 (1913).

[95] *Id.,* at 60.

[96] *Oregon-Washington R. & Nav. Co.* v. *State of Washington,* 270 U.S. 84, 46 S. Ct. 279 (1926), reversing *State* v. *Oregon-Washington R. & Nav. Co.,* 128 Wash. 365, 223 P. 600 (1924).

[97] *Asbell* v. *State of Kansas,* 209 U.S. 251, 28 S. Ct. 485 (1908) affirming *State* v. *Asbell,* 74 Kan. 397 (1906); also unhealthy swine or cattle *Mintz* v. *Baldwin,* 289 U.S. 346); decayed or noxious foods (*Grossman* v. *Lurman,* 192 U.S. 189); dead bodies of animals not slaughtered for food (*Bowman* v. *Chicago & N.W. Railroad Co.,* 125 U.S. 465).

[98] *Clason* v. *Indiana,* 306 U.S. 439 (1939)

99   "The New Jersey statute is not such a quarantine law. There has been no claim here that the very movement of waste into or through New Jersey endangers health, or that waste must be disposed of as soon and as close to its point of generation as possible. The harms caused by waste are said to arise after its disposal in landfill sites, and at that point, as New Jersey concedes, there is no basis to distinguish out-of-state waste from domestic waste. If one is inherently harmful, so is the other." 437 U.S. 617 (1978). See also *United States* v. *Pennsylvania Refuse Ass'n.,* 357 F. 2d. 806 (3rd Cir. 1966), Cert. den. 384 U.S. 961 (1966), "...refuse transported from Pennsylvania to New Jersey and disposed of in the latter site...was plainly a proper subject of interstate commerce" (at 808).

100   "I simply see no way to distinguish solid waste, on the record of this case, from germ-infected rags, diseased meat, and other noxious items...Because I find no rationale basis for distinguishing the laws under challenge here from our past cases upholding state laws that prohibit the importation of items that could endanger the population of the state, I dissent." (Rehnquist, J.)

101   *City of Philadelphia* at 625.

102   *Pike* v. *Bruce Church, Inc.,* 397 U.S. 137 (1970.)

103   *Huron Portland Cement Co.* v. *Detroit,* 362 U.S. 440 (1960).

104   *Huron Portland Cement Co.,* at 445.

105   *Id.,* at 442.

106   *The Great Atlantic and Pacific Tea Company, Inc.* v. *Cottrell,* 424 U.S. 366, 96 S. Ct. 923 (1976), *Polar Ice Cream & Creamery Co.* v. *Andrews,* 375 U.S. 361, 84 S. Ct. 378 (1964), *H.P. Hood & Sons, Inc.* v. *DuMond,* 336 U.S. 525, 69 S. Ct. 657 (1949), *Dean Milk Co.* v. *Madison,* 340 U.S. 349, 71 S. Ct. 295 (1951), *Milk Control Board* v. *Eisenberg Farm Products,* 306 U.S. 346, 59 S. Ct. 528 (1939), *Baldwin* v. *G.A. Seelig, Inc.,* 294 U.S. 511, 55 S. Ct. 497 (1935). "The mandatory reciprocity provision of Sec. 11, insofar as justified by the State as an economic measure, is precisely the kind of hindrance to the introduction of milk from other states...condemned as an unreasonable clog upon the mobility of commerce...[It is] hostile in conception as well as burdensome in result." *Great Atlantic & Pacific Tea Co., Inc.* at 380, citing *Polar Ice Cream & Creamery Co., supra.*

107   437 U.S. 617 (1978).

108   *Id.,* at 622.

109   *Hunt* v. *Washington Apple Advertising Comm'n.,* 432 U.S. 333 (1977).

110   *City of Philadelphia* at

111   *Hardage* v. *Atkins,* 619 F.2d. 871 (1980).

112   That statute held in *Hardage* to violate the commerce clause was Section 2764 of Oklahoma statute title 63 (Supp. 1973), which provided: "The [Controlled Industrial Waste Management Section] shall disapprove any plan which entails the shipping of controlled industrial waste into the State of Oklahoma, unless the state or origin has enacted substantially similar standards for controlled industrial waste disposal as, and has entered into a reciprocity agreement with, the State of Oklahoma. The determination as to whether or not the state or origin has substantially similar standards for controlled industrial waste disposal is to be made by the Director of the [Controlled Industrial Waste Management Section], and all reciprocity agreements must be approved and signed by the Governor of Oklahoma."

113   Our conclusion is then that the "substantially similar standards" provision is not different from the required reciprocity agreement. Both constitute a mandatory scheme which violates the Commerce Clause. The "similar standards" requirement is within the scope of the decisions of the Supreme Court.

114   See generally, Higgins, Oklahoma Industrial Waste Statute Held Unconstitutional, 21 Nat. R. J. 185 (1981). Also see *Reeves* v. *Kelley,* 603 F. 2d 736 (8th Cir. 1979), *Washington State Bldg. & Constru. Trades Council* v. *Spellman,* 518 F. Supp. 928 (E.D. Wash. 1981) (overturning a state initiative that prohibited a disposal facility from accepting nuclear waste produced from out-of-state; *Brown-Ferris, Inc.* v. *Anne Arundel County,* 438 A. 2d 269 (Md. 1981) (invalidating an ordinance prohibiting disposal and transporting through the county of wastes originally out-of-state).

115   *One-Way Disposal, Inc.* v. *City of Portland,* 14 E.R.C. 1577 (1980).

116   393 U.S. 129 (1968).

117   At 140.

118   At 630-631.

119   330 F. Supp. 1218 (S.D. Fla. 1971).

Chapter 3

REMEDIAL OPTIONS IN HAZARDOUS WASTE MANAGEMENT — AN OVERVIEW*

Edward W. Kleppinger

TABLE OF CONTENTS

---

* All footnotes appear at end of chapter.

I. INTRODUCTION

There is a limited set of fairly simple strategies which may be utilized in managing a hazardous waste problem once it begins to deteriorate and enforcement action becomes increasingly imminent. This is possible, notwithstanding the observation that each potential case is factually specific and thus unique. These are potential cases because, as discussed in some detail below, some of the best defensive actions can be taken before a situation results in litigation. At a minimum, these strategies enhance preparation for litigation and may result in enforcement actions being abandoned before complaints are ever filed.

Management of actions of these types is rapidly evolving. Typically, what has happened in hazardous waste enforcement actions is that lawyers without significant technical expertise or experience defend these cases using very specialized technical experts. In general, technical experts are quite capable of defining the degree of toxicity of a particular chemical agent, but have difficulty relating the bigger picture of the ultimate health effect and cost benefit issues.

II. REMEDIAL OPTIONS

A. The Basic Strategies

While there are a variety of technologies which can be utilized to handle hazardous waste situations, the remedial strategies which are available in any particular situation may be summarized by a simple three-by-two matrix. This results in essentially six combinations of remedial actions. Remedial actions may be accomplished either at the site of contamination or off-site. Whether on-site or off-site, three separate technical strategic options are generally available. One option is to allow leakage of the waste products into the general environment. Containment is also an option; that is to say, the hazardous wastes are contained in some fashion in isolation from the general environment. The third option is to detoxify, treat, or destroy the hazardous materials. Incineration is the best known example of treatment.

These basic six strategic options are all that are available, although there are combinations of these options. For example, the containment strategy generally generates a significant volume of contaminated leachate which must be handled. While historically some states (e.g., California) have allowed the leachate to be placed back in the containment system, more modern practice is to require the leachate to be treated and discharged or destroyed by incineration. The difference between a containment strategy and a leakage strategy is one of degree, not one of absolute difference, since there is no such thing as total containment.

The above point about containment strategies is an important one. "Impermeable" clays still have a degree of permeability however low. Large landfills with a very low permeability still allow a significant amount of leakage of rain water into the fill, thus creating significant volumes of leachate. It has been argued that synthetic liners have an absolute impermeability, but this has been found not to be true.[1]

At the present time, containment strategies are popular. The reason for this is that they are generally perceived to be less expensive than treatment strategies. For this reason, most hazardous waste remedial programs which have been implemented to date have featured containment approaches. The best example of this type of approach is Hooker Chemical Company's settlement with the Michigan Department of Natural Resources at its Montague, Michigan site which resulted in Hooker constructing a secure clay vault with leachate collection to contain a series of waste materials. (*Kelly et al.* v. *Hooker Chemical & Plastics Corp.* No. 79-22878-CE). The EPA and New York State Department of Environmental Conservation have also emphasized what they classify as a containment strategy in the settlement of Hooker's waste dump at Bloody Run (Hyde Park) Dumpsite in Niagara Falls (W.D.N.Y. 79-989).[2]

Containment strategies are characterized by low capital costs, but high maintenance and operational costs. They require devices which must be monitored and maintained for the life of the hazard. Furthermore, insurance must be purchased for the life of the facility. This methodology is contrasted with treatment procedures, such as incineration, which have high start-up costs. The essential advantage with an effective treatment procedure is that the hazardous material is effectively destroyed with virtually no future costs.

The future costs for perpetual care and monitoring for containment strategies are generally not, as yet, fully appreciated. These costs, if discounted to present value and adjusted for inflation rates, can be highly significant.[3] This was the conclusion reached by Mead Corporation in a preliminary analysis of four of Hooker's dumpsites in Niagara Falls. In preparing these figures to ascertain potential financial liability for Hooker's parent, Occidental Petroleum Corporation, in a corporate takeover attempt, it was determined that the "total" costs of containment approaches used by Hooker were of the same order of magnitude as the cost for incineration for each of these dumpsites. In short, the present popularity of the containment strategies may be to a substantial extent as a result of a perceived lower cost.

It is critical to evaluate each of these strategies against the facts of the particular hazardous waste situation. This evaluation should give careful consideration to the downside risk of each of these strategies as well as the advantages. For example, one questions the efficacy of selecting an on-site containment strategy as Hooker did in its Bloody Run Landfill, when the basic geology and the history of depositing waste directly on fractured bedrock largely precludes containment over the life of the hazard.

B. On-site vs. Off-site

The fundamental difference between on-site and off-site treatment is primarily one of citizen and public involvement. If the public becomes intimately involved and the site becomes subject to media attention, then typically on-site remedial options will need to be implemented, since public pressure rarely allows off-site options to be utilized. For example, in attempting to determine what costs Hooker Chemical might be liable for in off-site disposal of waste at Montague, a series of disposal sites were contacted. The disposal sites in each case gave quotes for disposal of typical waste generated. However, once the potential source of the waste was identified, the disposal cost often doubled or tripled, that is, if the site operator was still willing to handle the waste. There was no change in the waste characteristics but only in the site operator's perception of public reaction to the specific waste disposal and the implied threat to the existence of his disposal operation. The mere suggestion that Hooker waste from Swartz Creek, Michigan might be disposed at a Detroit area secure landfill led the sheriff deputies to set up road blocks into the disposal site itself. The waste was eventually hauled back to Hooker's Montague site for containment.

III. DEFINING CLEAN

A. Establishing an Acceptable Standard

One area concerning implementing remedial options is the question of how clean is clean and how to define clean. The public and regulatory authorities will generally adopt a position briefly summarized as, "well, clean it all up." This is fairly simple and a tenable solution if there is a single drum of material located in a field somewhere and the drum has yet to leak.

Typically, the next approach on the part of regulatory officials is to suggest that it be cleaned up to "background" levels. This assumes that there is a known background level of the particular material or materials in question. With the exotic nature of certain hazardous wastes, this assumption is not always true.

It may be generally stated that if you leave it to the regulatory agencies to suggest an appropriate standard, you may well end up with a level that represents absolute pure background, generally the equivalent in trying to live with the Delaney Amendment, where analytical techniques are constantly improving and certain substances can be detected at the part per trillion level or lower. The "how clean is clean" argument also develops when one attempts to implement a strategy of allowing slow leakage to the environment or one with less than total containment.

Where ground-water contamination is purging into a surface stream and causing minimal environmental impacts, arguments of risk assessment have been successfully used. Regulatory agencies have often taken the position that the discharge constitutes an uncontrolled discharge of pollutants and therefore must be controlled, often without adequate regard to the context of the discharge, the amount of material, or the type of material. Counter arguments include the use of a risk assessment. An example of risk assessment is the computation of the impact of a known carcinogen in the discharge for its potential for causing cancer. Once the risk is determined, that risk is then compared with the risk presented by common circumstances. In a Michigan case, for example, it was satisfactorily shown that the risk created by seepage of halocarbons into a stream was actually two to three orders of magnitude less than the present risk of drinking the city water supply.

Another excellent defense is to attempt to determine other discharges which have been controlled but which contain the same or similar compounds. In this strategy, it is argued that there are still significant amounts of chemicals discharged to the environment even though a high degree of control exists. Information on Natural Pollutant Discharge Elimination System permits and samplings available under Freedom of Information Act requests help establish this. The argument is made that if the regulatory agency has approved the continuing discharge of 1 lb of a particular chemical after it has been treated, then the "uncontrolled" discharge of a tenth of a pound of that same chemical in a similar environment should not be found to be objectionable.

In dealing with hazardous waste, it is sometimes forgotten that many of the components of hazardous waste are present in commercial products used every day. If the housewife knew what was under her sink in terms of hazardous materials, she probably would be scared to death. Thus, in one particular situation it was shown that the ethylenedichloride emissions from residual contamination were minimal when compared to the emissions of ethylenedichloride occuring while filling an auto gasoline tank. In establishing clean levels for various compounds, it is useful to determine where those compounds are utilized in commercial products and what kinds of exposures therefore occur.

The Federal Government has been in the oil and hazardous material business for some time now. While there are differences between hazardous waste lawsuits and the clean-up of a spill of hazardous materials, there are also some useful similarities, particularly in regard to establishing a meaningful standard. One should investigate to determine what levels have been achieved by the government in a spill clean-up program. This can help set the levels which should be proposed in a hazardous waste enforcement case. This has been a particularly difficult problem for the government in revising the National Contingency Plan as required by Superfund.[4] While the EPAs public position is that the extent of clean-up should be the same whether it is financed under Superfund or financed privately, this causes some problems when the Section 104 cost balancing test is utilized. That is to say that if a clean-up must be cost effective, then this itself begins to dictate the level to which contamination is cleaned up.

B. Cost Effectiveness

The government's approach to cost effectiveness is to determine the most cost-effective way to meet a particular background level. It has not been approached from the perspective of which background level is it cost effective to clean up. The argument is available in litigation since the actions are generally equity actions and involve a balancing of the equities, which the technician can translate as cost-effective solutions.

One aspect of this issue which has received little attention is that of permanent abandonment of ground water resources. While this sounds somewhat heretical since half the water utilized in the U.S. is ground water, it may, in fact, be the best approach. Over 4 years ago, then Representative Toby Moffet (Dem. Conn.) observed that certain ground-water resources might have to be abandoned because of their pollution. At the Ott/Story/Codova site in Michigan, it was determined that even with an extensive ground-water purging program costing many millions of dollars, the ground-water resource could no longer be utilized to supply drinking water. If the ground-water resource is abandoned as a drinking water supply and no amount of clean-up will restore it, then one questions the necessity of any substantial clean-up. This is particularly true in the Ott case, when the plume of contamination has been stabilized.

IV. ALTERNATE DEFENSIVE STRATEGIES

There are two general strategies in dealing with hazardous waste litigation. These are generally classified as either passive or aggressive.

The problem rarely goes away with respect to hazardous waste, it can only get worse. As ground-water contamination occurs and increasing amounts of ground-water become contaminated, it is self-evident that the cost of clean-up to any particular background level also becomes increasingly expensive. It is a basic law of thermodynamics that much more concentrated materials are easier and cheaper to handle than dilute materials.

A. The Passive Strategy

The passive strategy allows organized opposition to develop. It tends to develop long delays as no one wants to make a decision and more people must be involved in a decision train. It tends to engender third party suits and it can remove the clean-up from private hands and put it into government contractor hands. Government contractors have little incentive to minimize costs since they are typically paid on a cost plus fixed fee basis. The downside risk to a passive strategy is that the site may become the subject of considerable media attention. Furthermore, the solutions ultimately reached and the levels of cleanliness required stand a substantial chance of being, in the final analysis, more expensive. Other intangible damage may be done; for example, loss of a good corporate name. Furthermore, focus may be placed on other sites owned by the corporation. To summarize, the problem with the passive approach is that it can result in the most expensive solution being implemented ($26 million so far at Chemical Control Corp — say, $1000 per drum). The danger of the passive approach is perhaps best illustrated by Hooker Chemical Company's experience at Bloody Run.

There may be cases where a passive strategy is correct. The first involves a situation where a generator has minimal contribution to an off-site commercial disposal site. In this case, hopefully other generators will be identified and the minor nature of the contribution can be argued in an equity court. Even in this case, it may be appropriate to exercise an aggressive strategy in going out and identifying other generators for the government regulatory agencies so that they may also be sued for the clean-up costs. The other situation where a passive strategy may be appropriate is that of the multi-generator lawsuit wherein a site is proposed for clean-up in which a series of generators have deposited waste. The lawyers in these cases talk about an "indivisible harm" situation. Presently, EPA has proposed and the Department of Justice has concurred with a joint and several liability theory to attack this particular problem. This may be best defended in court by arguing the equities involved.

There are avenues which should be explored to minimize the exposure to the downside risk of a passive wait and see strategy. For example, minimally, the generator should attempt to see whether waste generated by the company can be identified and retrieved. It may be possible to buy out the situation early on at a minimum cost as opposed to fighting it and taking the potential downside risk of an adverse judgment. This is the so-called "pay and walk" strategy. See Seymour Recycling (Civ. No. IP-80-457-C.D.C.S.D. Ind.). Even if a particular waste cannot be identified as being contributed by the company, perhaps waste can be removed which can be identified as similar in kind and quantity.

It may be useful to bind the generator group into a prejudgment agreement so as to be able to exercise an aggressive strategy, thus removing the clean-up from government hands.

B. Aggressive Strategy

The aggressive strategy is largely self explanatory. If you control the situation, other people cannot. As noted, one advantage of employing an aggressive strategy is the observation that hazardous waste situations typically only get worse and more expensive to clean up as time goes on. As public involvement increases, the flexibility in achieving solutions decreases, eventually converging on the most expensive solution. The power of the aggressive approach lies in assuming the white hat of the "good guy". Thus if one establishes an aggressive approach and in fact takes the lead in implementing a solution to the problem which is generally cost effective and environmentally sound, then the regulatory agency may be placed in a role of saying "no" to an ongoing, environmentally sound solution. Furthermore, while wearing the white hat, the blame can be spread around to others. For example, at most sites it may be argued that the government regulatory agency itself failed in its duty to monitor or inspect the hazardous waste disposal site. The aggressive approach minimized the chance for public concerns to find expression in either rational or irrational attacks.

To illustrate how the aggressive approach can work, two examples come to mind. One has not involved EPA litigation and probably will not, although if it had been allowed to proceed with a passive approach, litigation most likely would have commenced. The other case described did involve a RCRA §7003 action.

Production at the Ott/Story/Codova site in Michigan started in approximately 1957. Wastes were put into seepage lagoons under state permit. The lagoons caused extensive ground-water contamination which initially contaminated the plants own water supply. This original company was purchased by a large corporation through an independent subsidiary in 1965. This particular subsidiary then proceeded to spend several millions of dollars installing equipment with a twofold objective of eliminating the source of ground-water contamination and containing and controlling the existing ground-water contamination. Both an incinerator with a venturi scrubber and purge wells were installed. Waste-water treatment began and surface discharges replaced seepage lagoons.

In 1972, the subsidiary sold its assets and liabilities with clear disclosure of the perceived environmental problems to a new company called Story Chemical Corporation. Officers of the Story Chemical Corporation then proceeded to turn off the purgewell systems. While proceeding to accumulate significant volumes of hazardous waste onsite within months, off-site potable wells became contaminated. Story Chemical Corporation then filed for bankruptcy in 1976 and was adjudged bankrupt in 1977.

The assets of Story Chemical were purchased at a large discount by Codova Chemical Company, a subsidiary of Aerojet General. As part of this purchase, Codova negotiated with the State of Michigan Department of Natural Resources to fund, along with a state appropriation, the clean-up of sludges and several thousand barrels of waste and to purge the ground water. Cordova also agreed to supply an alternate water supply to citizens whose wells had been impacted. In 1978, the State Department of Natural Resources eventually did clean up the concentrated waste, but failed to install a ground-water purge system of potable water for the residents. In the meantime, a civil suit class action was filed, not against the bankrupt Story Chemical, but against the parent of the independent subsidiary. This lawsuit was proceeding slowly down the legal chain of events.

A passive defense might have involved relying totally on successful defense of the class action. Suffice it to say that the corporation and its lawyers felt that they had significant defenses in this regard. For example, a serious question existed whether the suit could pierce the corporate veil. However, rather than exercise this particular defense, the company involved chose to manage the situation in a very aggressive way. The strategy was simple; it was decided to resolve the potential public health issue first with the idea of ultimately resolving the environmental issues, if necessary, afterward. In other words, the contamination at this site was separated into public health issues and environmental contamination issues. With respect to resolving the public health issue involving potable drinking water, it was decided to implement a public water supply system. The class action lawsuit was used as a vehicle for implementing the solution and necessary technical studies were funded to demonstrate that it was an economically as well as an environmentally sound solution. Eventually, the county agreed to own and operate the system which was proposed as one of two alternatives. The alternative chosen by the county was the extension of a water main to the area serving not only those directly impacted (six to nine houses) but several thousand people who were not and could not have been impacted by this particular ground-water contamination issue.

With the resolution of any potential public health issues, environmental constraints can often be rationally addressed. With the 1978 removal of concentrated materials only two questions remained; the contaminated ground water and the residual ground contamination. Since the ground-water plume was stable because it was purging into a short stretch of a small surface water steam and since the ground contamination had existed in a high water table area for some time, it was agreed that the controlling factor in deciding what to do was an analysis of what the impact was on the small stream. The Michigan Department of Natural Resources analyzed the impact at $13,000/year. This could immediately be set against the DNR's proposed purging plan which involved some $30 million in costs.

The point to be made is that there are benefits, if one is willing and able to get out in front and control the flow of events in these cases in moving towards an economic, but environmentally rational solution to the problem. Even if the plan had failed, the company would have other legal defenses upon which to rely.

An example of an enforcement action had already been brought is Metal Bank of America, Inc., which is located in North Philadelphia, Pa. The site involved an operation which reclaimed copper cores from transformers. Some of the transformers, it was alleged, contained PCBs which were not drained and the oils from these transformers supposedly escaped from a tank which was to contain them. Thus, there was a layer of oil potentially contaminated with PCBs which existed in fill material along the banks of the Delaware River. The Coast Guard had brought an oil spill action and that had been resolved, but EPA and the Department of Justice brought a RCRA §7003 suit in an attempt to force further clean-up of the site.

EPA's proposed control plan involved recovering the oil using sheet pile around the site to seal off the site from the river and then excavating all contaminated material with removal to a PCB disposal site followed by back-filling with clean fill and a clay cap. Counter-pumping was also proposed.[5]

In this case, an aggressive strategy was also employed. A remedial program was carefully installed which skirted EPA's permitting requirements. This proceeded independently of the lawsuit. Thus, while the lawsuit was moving along with discovery, etc., this remedial program was being utilized to retrieve the recoverable PCBs, leaving in place the nonmobile PCBs.

This particular aggressive strategy was so successful that the U.S. Attorney and Regional Attorney for EPA have agreed to accept a stipulated settlement rather than a decree of consent.

V. SUMMARY

The hazardous waste issue is not going to go away. The congressional pressure is increasing. Most congressmen have a waste site in their district. Polls show hazardous waste very high in the general public's perception of problems facing the U.S. Furthermore, if one has a perception that EPA is going to back away from this issue, one simply has to examine EPA budget requests to the Office of Management and Budget for the enforcement effort in hazardous waste.

There is an essential difference between hazardous waste problems and those air and water pollution problems which technical consultants and corporate management have been used to dealing with. In the case of air and water pollution problems, the strategy of waiting typically was not as economically or politically damaging as it is in the case of hazardous waste. This is because in the case of an air pollution problem, once the air pollution control equipment is installed, the problem goes away. As the regulatory agency develops and implements its regulations, at some point the corporation complies whether they comply quickly or whether they comply after losing a series of lawsuits, they comply; either they install the air pollution control equipment or if economically that is not viable, they close the facility or close the operation generating the air pollution. The same is true of most water problems. The problem then goes away.

In case of hazardous waste, this is not true. Suppose a particular company is generating still bottom which is highly hazardous. Let us take, for example, the trichlorophenol still bottoms that Hooker Chemical generated. Eventually, Hooker Chemical eliminated this process. Dow Chemical has not, but has incinerated these still bottoms for a number of years, thus eliminating them from the environment. Now in the case of an air or water pollution problem, once the process was abandoned or once incineration has taken place, then the problem disappeared. With hazardous waste, this is not true because the trichlorophenol still bottoms that Hooker generated are still there to haunt them. There is over a ton of dioxin from trichlorophenol still bottoms in Bloody Run alone. The threat and the potential economic loss imposed controlling this material did not go away when Hooker stopped the process generating the material.

The problem with hazardous waste from past disposal practices can only get worse, and thus it is best to get in and aggressively manage these issues rather than wait for the inevitable.

Another essential difference between the control of air and water pollution is that, with respect to hazardous waste, paranoia has set in. Every newspaper article starts off with "our town's Love Canal". Part of the problem involves defining just what a hazard is. Why, for example, if the ethylenedichloride and benezene in gasoline are not a perceived hazard, does the drum of ethylenedichloride sitting at a defunct disposal site suddenly become a horrifying hazard which can potentially wipe out a community and cause congressmen to call to arms?

With respect to hazardous waste, there is a $1.6 billion potential fund set up to clean these sites. It is clear that regulatory agencies and congressmen are very interested in "getting their share". Thus, a company waiting for the government to clean it up and then getting fixed with liability built into CERCLA probably faces at least a ten times liability cost due to the fact that when the government cleans it up, it chooses the most expensive, safe, secure system for doing so without significant attention to the larger issues of cost effectiveness.

It is generally best to manage the problem aggressively before it manages you. Managing the problem aggressively requires the close cooperation of the legal profession and those who have *general* technical expertise in the various technical disciplines required to analyze and resolve hazardous waste problems.

FOOTNOTES

[1] See Peter Montague, Ph.D., Princeton University, "Four Secure Land Fills in New Jersey, a Study of the State of the Art", draft dated July 19, 1981.

[2] Third parties have argued that the containment strategy selected at Bloody Run is not really a containment strategy since the container which is to be constructed by Hooker Chemical does not contain a "bottom". The bottom of the landfill is fractured bedrock which is allegedly contaminated for a depth of hundreds of feet.

[3] For further information, see *The Comparative Evaluations of Incinerator and Landfills for Hazardous Waste Management* prepared for Chemical Manufacturer Association by Engineering-Science, May 25, 1982. Landfill at one third the cost of incineration is the conclusion of this study greatly favoring landfills. Also see *Chemical Waste Disposal Facility Study: Heard County, Georgia,* John L. Carden, Center for Environmental Safety, Georgia Institute of Technology (February 9, 1981) projecting the costs for perpetual care and maintenance expenses at 10% inflation as being $13.2 million for the first 30 years and $154 million for 200 years total life of facility.

[4] EPA makes a distinction between cost effective — the best way to achieve a level, and cost benefit — the cost of the benefits achieved.

[5] As a side light, we were able to show that statistically, EPA's control plan would kill at least two people in accidents due to the truck ton mileage which would be involved in hauling this material off-site based on accident statistics. Not all risk assessments need to be highly detailed.

Chapter 4

FROM CRADLE TO GRAVE: THE LEGISLATIVE HISTORY OF RCRA

Jeffrey H. Howard*

TABLE OF CONTENTS

---

* Jeffrey Howard, Esq., Davis, Graham & Stubbs, Washington, D.C.

On October 21, 1976, the Resource Conservation and Recovery Act of 1976 (RCRA, Pub. L. 94-580,90 Stat. 2795) became the law of the land. The declared purpose of this comprehensive, new legislation was to close the last loophole in environmental regulation by regulating hazardous waste "from cradle to grave". RCRA's legislative history from its early development through its 1980 Amendments and the most recent amendments in 1984 is not only important to an understanding of the statute itself, but also serves as an example of contemporary environmental law-making where the legislative goals are often broad and the "real world" objectives are specific, precise, and intermixed with many practical problems of analytical chemistry, toxicology, and environmental engineering.

The following discussion traces the legislative history of the 1976 law, describes the more recent 1980 and 1984 statutory amendments, and concludes by highlighting significant issues of legislative intent which remain to be resolved by Congress or the Courts.

I. THE 1976 LAW

A. Beginnings of Hazardous Waste Policy

The origin of Federal hazardous waste policy dates back to the Public Health Service Act of the early 1950s (see S. Rep. No. 988, 94th Cong., 2d Sess. 5-6 (1976)). The Public Health Service program would hardly qualify as a hazardous waste policy by today's standards, but it represented the Federal Government's first attempt to stimulate state and local improvements in solid waste storage, collection, and disposal with an emphasis on waste disposal research at the federal level. The results of program surveys in 1965 evidenced only 12 states reporting identifiable solid waste activities and 31 states with no such program whatsoever.

The next step was the Solid Waste Disposal Act of 1965, Pub. L. 89-272, 79 Stat. 997 (1965) (SWDA), which represents the first significant effort to set forth a national plan for solid waste management. Actually passed as Title II of the Clean Air Act, Pub. L. 89-272, 79 Stat. 995 (1965), SWDA had two major purposes; to initiate research and development into methods of solid waste disposal and conservation by reducing waste through recovery and utilization; and to provide technical and financial assistance to state and local governments and interstate agencies for planning, development, and implementation of disposal programs and surveys of waste practices. Congress intended SWDA to establish a mechanism for long-range research and to foster interlocal cooperation in attacking waste problems. Although SWDA broadened the federal role through research plans and extended financial assistance, and provided the Department of Health, Education and Welfare and the Department of Interior with additional functions, the Act did not provide any regulatory powers with respect to solid waste.

It was not until 1970 that Congress began to turn its attention to the type of wastes which are now referred to as "hazardous" and established a policy which was slightly more regulatory in nature. The Resource Recovery Act of 1970, Pub. L. 91-512, 84 Stat. 1228 (1970) (RRA), marked a policy shift away from disposal and toward maximum recovery of reusable materials and energy from solid waste. The RRA also increased the emphasis on management to reduce the volume of waste.

The RRA contained recommendations to accomplish these goals of recovery and management which were not contained in its predecessor SWDA. Research and development policies were broadened to include support of major demonstrations (including grants and technical assistance), applying solid waste management and resource recovery systems to "preserve and enhance the quality of air, water, and land resources." RRA § 202 (b)(1). The Act attempted to broaden federal support in other areas including: planning and developing state programs; providing federal training grants in occupations involving the design, operation, and maintenance of solid waste recovery systems; and funding comprehensive studies on solid waste management practices.

In the RRA, Congress also attempted to impose stronger federal support for recovery and reuse with a new element — agency guidelines — perhaps foreshadowing the comprehensive regulatory program to come. These "guidelines" on waste management, issued by the Secretary of HEW (later to come under EPA jurisdiction), would be mandatory for federal agencies, advisory to others, and should "be consistent with public health and welfare, and air and water quality standards and adaptable to appropriate land use plans." RRA, § 209(a). The Secretary was also to recommend model codes, ordinances, and statutes. The Act increased agency solid waste responsibilities in two major ways: first, HEW was to study and develop methods of reducing waste volume through both reduction of waste generation and increased recovery by studies of packaging practices, incentives, and disincentives for reclamation, and effects of present policies; second, HEW was to provide financial and technical assistance to states for construction of projects utilizing new technologies for reducing the environmental impact of waste disposal and promoting recovery.

Spurred by the RRA, federal solid waste programs began appearing in the early 1970s at the Energy Research and Development Administration, the Bureau of Mines, the Federal Energy Administration, the National Science Foundation, the National Aeronautics and Space Administration, the Housing and Urban Development Department, and the Tennessee Valley Authority.

Despite these federal programs and the RRA's emphasis on research and recovery, federal efforts in solid waste management moved slowly during the early 1970s, and Congress later viewed this period as a time when solid waste programs were actually set back because of executive branch policies which included administrative reorganization, delayed submittals of reports to Congress, and reduced budget requests. S. Rep. No. 988, 94th Cong., 2d Sess. 6 (1976).

Perhaps reflecting this uncertainty, in 1973, Congress was unable to conduct appropriate oversight hearings on solid waste management and simply adopted a "One Year Extension of the Solid Waste Disposal Act", Pub. L. 93-14, 87 Stat. 11 (1973), which deferred further legislative consideration. In order to develop legislative proposals and hold appropriate oversight hearings, the Senate Public Works Committee established the Panel on Materials Policy. This Panel began extensive hearings in June and July of 1974, considering an Administration bill and several additional bills.

B. Three Major Bills

The first Bill to provide for the regulation of hazardous waste was S. 1086, the "Hazardous Waste Management Act of 1973", introduced on behalf of the Nixon Administration by Senator Howard Baker on March 6, 1973. 119 *Cong. Rec.* 6435 (1973). Senator Baker's Bill would have defined "waste" as "useless, unwanted, or discarded solid, semisolid, or liquid materials." S. 1086, § 3(3). The term "hazardous" was defined in terms similar to those used in RCRA Section 1004(5)(A) and (B), but the Bill omitted the RCRA Section 1004(5)(A) reference to wastes which cause an increase in mortality or illness, as an alternative ground for classifying as hazardous.

S. 1086 was remarkably similar to the ultimate version of RCRA adopted 3 years later. The Nixon Administration Bill called for EPA to identify hazardous wastes, adopt standards for generators and for treatment and disposal facilities, issue permits for treatment and disposal facilities which met EPA's standards, enforce the statute by Administrative order and a $25,000 per day civil penalty for continued violation, and it prohibited unpermitted hazardous waste disposal. In addition, S. 1086 would have authorized EPA to obtain a District Court injunction against any facility that "presents an imminent and substantial danger to human health or the environment."

However, S. 1086 differs from RCRA in several important respects: (1) it did not regulate the transportation of hazardous wastes or require a manifest system for such wastes; (2) it expressly directed that in establishing hazardous waste standards, EPA "shall take into account the economic and social costs and benefits of achieving such standards"; (3) it did not provide for any regulation of nonhazardous solid wastes such as the state-enforced controls, provided in RCRA Subtitle D, over "open dumping"; and (4) neither the definition of "hazardous waste" nor any other provision exempted other environmental controls from coverage by the Bill (such as wastes discharged by an NPDES facility), nor would it have provided for any special treatment for mining or oil and gas wastes.

The 1973 Nixon Administration Bill apparently got lost in the shuffle of Watergate and no significant action was taken on the matter for 2 years. Even later, in April 1975, Deputy Administrator Quarles testified that EPA had been directed by President Ford's Office of Management and Budget "to hold the line against all new spending programs in order to fight inflation and keep the budget in control" and thus the Ford Administration did not support the earlier Nixon Administration Bill (Waste Control Act of 1975: Hearings on H.R. 5487 Before the Subcommittee on Transportation and Commerce of the House Committee on Interstate and Foreign Commerce, 94th Cong., 1st Sess. 759 [1975]. Several other relatively minor bills dealing with hazardous waste were introduced in 1974, but none developed beyond the Committee stage. See, e.g., S. Rep. No. 1127, 93rd Cong., 2d Sess. 7-9, 16-17, 43, 45 (1974).

On July 21, 1975, Senators Randolph and Hart introduced S. 2150, the second comprehensive bill to regulate hazardous waste. Hearings were held by the Senate Public Works Committee and the Bill was soon ready to be reported out of Committee. However, because an authorization Bill needed to be reported out by May 15 in order to comply with the Congressional Budget Act, the Senate Public Works Committee initially reported S. 2150 as a simple reauthorization bill only. See S. Rep. No. 869, 94th Cong., 2d Sess. (May 13, 1976). One month later, the Senate Committee staff had completed a report on the substantive hazardous waste bill and it was then reported as a proposed amendment to S. 2150. See S. Rep. No. 988, 94th Cong., 2d Sess. (1976). Senator Randolph then offered the Committee's Bill as an amendment to the authorization bill, S. 2150. See 122 *Cong. Rec.* 20733 (1976). A few days later, after rejecting a mandatory "returnable bottle" amendment offered by Senator Hatfield (122 *Cong. Rec.* 21404, 21417 [1976]) and agreeing to a compromise provision calling for a study on returnable bottles (*id.,* at 21419), on June 30, 1976 the Senate adopted S. 2150 (*id.,* at 21429).

S. 2150 built upon the 1973 Nixon Administration Bill and added several important features. The Bill directed EPA to establish criteria for identifying hazardous waste and guidelines for determining harmful quantities of such wastes. EPA was to publish a list of wastes which were hazardous according to the criteria. In doing so, the lists established under Sections 307 and 311 of the Federal Water Pollution Control Act and Section 112 of the Clean Air Act were to be included. Twelve months after promulgation of the list, disposal of such wastes would be prohibited except in accordance with a permit. Permits could be issued for treatment, storage, or disposal on the condition that there would be "no disposal of any designated hazardous waste in harmful quantities." S. 2150, § 212(b)(1).

Although the Senate Bill established no direct controls over generators and transporters of hazardous waste, it did provide for a mandatory manifest system which was to originate with the generator, carry through the transporter, and continue through the person treating, storing, or disposing of the waste. S. 2150, § 212(b)(3). The states were to be authorized to issue such permits. Permits would be issued on the condition that specified reporting, waste management, containerization, transportation, contingency planning, training, licensing, and financial reponsibility requirements were satisfied. S. 2150, § 212(b)(2). In addition, EPA would have imminent hazard authority (S. 2150, § 213) similar to that which had been proposed in the earlier Nixon Administration Bill. Violations were punishable by civil penalties of up to $10,000/day, or criminal penalties of up to $25,000/day and/or 1 year imprisonment, or both. S. 2150 also set forth a basic framework for regulatory control over nonhazardous waste by means of a ban on so-called "open dumping", which EPA was authorized to define. S. 2150, § 209(a)(3), § 211.

However, S. 2150 did not authorize EPA to establish standards applicable to hazardous waste generators, transporters, or owners and operators of treatment, storage, or disposal facilities. The sole basis for regulatory control was to be the permit required for hazardous waste disposal. In addition, unlike the Nixon Administration's Bill, S. 2150 defined "solid waste" in a way which attempted to avoid coverage of domestic sewage, industrial point source discharges operating under NPDES permits and source, special nuclear, or by-product material as defined by the Atomic Energy Act of 1954. S. 2150, §6(a) at (4). This definition made its way into the final bill and into RCRA Section 1004(27).

Of some particular interest is the definition given in S. 2150 "hazardous waste":

> The term "hazardous waste" means a waste...which in the judgment of the Administrator may cause, or contribute to, an increase in mortality or an increase in serious irreversible, or incapacitating reversible, illness, taking into account the toxicity of such waste, its persistence, and degradability in nature, and its potential for accumulation or concentration in tissue, and other factors that may otherwise cause or contribute to adverse acute or chronic effects on the health of persons or other organisms. S. 2150, § 6(c) at (15).

This definition does not use the word "environment" and it seems principally devoted to human health hazards, rather than broader environmental concerns. One can only speculate that had this definition been adopted into law, it may have narrowed the scope of RCRA regulation by limiting the reach of the program to wastes with a higher degree of toxicity to humans.

In its brief report on the proposed hazardous waste regulatory program in S. 2150, the Senate Public Works Committee stated:

> The purpose of this section is to eliminate any disposal of a designated hazardous waste in locations or circumstances which might be harmful to the public health or the environment.

> The administrator is to promulgate criteria for identifying hazardous waste and issue guidelines which govern the determination in the permitting process whether a proposed practice or site will constitute disposal of a harmful quantity of hazardous waste. States are to use the criteria for identifying hazardous wastes for control under their programs. The Administrator also is required to publish a list of materials he determines to be hazardous wastes. In this initial list the Admistrator should concentrate on those hazardous wastes which pose the greatest threat to public health.

> This program is not a general regulatory program of municipal or private sanitary landfill operations. The only Federal requirement concerning general, non-hazardous solid wastes and landfills is the ban on open dumping in section 211. The permit program of section 212 is limited to designated hazardous wastes. It is not to be used to control the disposal of hazardous substances used in households or to extend control over general municipal wastes based on the presence of such substances. S. Rep. No. 988, 94th Cong., 2d Sess. 15-16 (1976).

In considering hazardous waste issues the House of Representatives was a few months behind the Senate. On September 9, 1976, the House Commerce Committee reported out H.R. 14496. See H.R. Rep. No. 1491, 94th Cong., 2d Sess. (1976). A few weeks later, on September 27, 1976, the House passed a modified version of H.R.

14496. The House Bill set the structural framework for RCRA and addressed the regulation of hazardous waste in a much more comprehensive manner than did the Senate
Bill.

Section 301 of H.R. 14496 would have required EPA to develop hazardous waste
criteria, characteristics, and lists. Unlike the Nixon Administration Bill (S. 1086) and
the Senate Bill (S. 2150), the House Bill (H.R. 14496, §§ 302-303) also authorized EPA
to establish specific standards applicable to generators and transporters of hazardous
waste apart from the permit and manifest systems. In addition to directly setting performance standards for treatment, storage, and disposal and requiring a permit for
such activities, the House Bill also provided for "interim status" — that is a grace
period during which facilities are allowed to continue operating while their RCRA
permit applications were pending.

In its report, the House Commerce Committee explained that in listing wastes as
hazardous, the Committee intended that EPA first adopt the hazard criteria which it
was then required to follow in listing particular wastes. The Committee believed that
this bifurcated regulatory process was preferable and that:

> The Committee anticipates the identification of two basic types of substances; those which
> are hazardous in their elemental and most common form, regardless of concentration, and
> those which when present in sufficient concentration or when mixed with other substances
> constitute hazardous waste.
>
> The criteria for identification of these substances should make such a distinction based on
> the danger to human health and the environment. The listing of any substance not found to
> be hazardous per se should be accompanied by an explanation as to when such wastes are
> considered hazardous. Such explanation should relate to the quantity, concentration, physical,
> chemical or infectious characteristics including toxicity, persistence and degradability in na-
> ture, potential for accumulation in human tissue and other factors such as flammability and
> corrosiveness which contribute to the hazardous nature of the substance, and which EPA is to
> consider as part of the listing process. H.R. Rep. No. 1491, 94th Cong., 2d Sess. at 25-26
> (1976).

With respect to EPA's authority under the Bill to control the activities of hazardous
waste generators, the House Report stated that:

> Rather, than place restrictions on the generation of hazardous waste, which in many in-
> stances would amount to interferences with the productive process itself, the Committee has
> limited the responsibility to the generator for hazardous waste to one of providing informa-
> tion.
>
> Through this process the Administrator will have at hand information on the location and
> volume of wastes being generated. Transporters of the wastes will know what the cargo con-
> tains, its general characteristics, and will have a warning as to its nature. Further, those who
> receive such wastes for treatment, storage, or disposal will have accurate knowledge of the
> characteristics and constituents of such waste prior to working with such wastes.
>
> Although there will be no requirement of the generator to modify his production process to
> reduce or eliminate the volume of hazardous waste, he will bear the burden of record-keeping,
> reporting to the Administrator, and providing information and warning to the transporter of
> the waste, and to those who treat, store or dispose of such wastes. *Id.,* at 26-27.

The manifest system called for by the House Bill was to become the unifying thread
of the entire regulatory program:

> It is not the Committee's intent to interfere with the transportation of the waste but rather
> to provide a system through which the movement of the waste can be traced. Too often trucks
> bearing hazardous waste have been unloaded along the roadside or a nearby landfill.
>
> The manifest system is intended to serve as a check against such practices. Originating with
> the generator, moving through the transportation stage, registered at an approved disposal site

for the treatment, storage or disposal of such hazardous waste and returned to the generator, the manifest will give to each party in the chain of handling a record.

In short, the duties of the transporter are to accept only those hazardous wastes properly labeled and in compliance with the manifest requirements, to discharge the specific duties of the transporter under the manifest system, and to deliver the hazardous waste only to the facility which the shipper designates on the manifest form to be a facility holding a permit issued under this title. *Id.,* at 27.

The House Committee plainly viewed the performance standards applicable to facilities which treat, store, or dispose of such wastes as the key to the program:

> The administrator is also required to promulgate performance standards applicable to those facilities operated for the treatment, storage, or disposal of wastes identified as hazardous. These performance standards must reasonably protect human health and the environment.

> The disposal facility requirements are the key provisions in the structure regulating the handling of hazardous waste. The manifest system finds its completion when such wastes are received by those who treat, store or dispose of such wastes and notice of receipt of such solid wastes are sent to the generator. . . .

> It is the intent of the Committee that responsibility for complying with the regulations pertaining to hazardous waste facilities rest equally with owners and operators of hazardous waste treatment, storage or disposal sites and facilities where the owner is not the operator. *Id.,* at 27-28.

C. The House - Senate Compromise

RCRA was adopted without a House-Senate Conference Committee to resolve differences between the House and Senate Bills. The Senate had adopted S. 2150 on June 30, 1976. 122 *Cong. Rec.* 21429 (1976). Subsequently, on September 9, 1976, the House Commerce Committee reported out H.R. 14496 (H.R. Rep. No. 1491, 94th Cong., 2d Sess. [1976]). As discussed above, the House Bill was much more comprehensive in its treatment of hazardous waste than the Senate Bill. Moreover, H.R. 14496, as reported out of Committee, enjoyed the support of industry, union, environmental, and state and local government groups. 122 *Cong. Rec.* 32598 (1976). In addition, the House Bill had been negotiated with the Office of Management and Budget and had the "full endorsement of the [Ford] Administration." *Id.* Negotiations with OMB during House Committee consideration had resulted in the elimination of a provision to create a United States Resource Recovery Corporation to stimulate resource recovery through low cost loans (former §§ 601-616 of H.R. 14496).

When the House Committee version of H.R. 14496 was initially considered on the floor, one of its managers, Congressman Skubitz, announced that there had already been negotiations between the Senate and House on the bill, and accordingly a compromise bill was introduced. As Congressman Skubitz explained:

> Mr. Chairman, this legislation has a companion bill in the Senate, S. 2150. The major differences between the two bills are, in my judgment, stylistic. Our legislation was written as a free-standing piece of new legislation. The Senate, for reasons of committee jurisdiction, amended the 1965 Solid Waste Disposal Act.

> Since we are now in the 11th hour of this Congress, we have decided to do all that is possible to avoid having to have a conference with the Senate. Therefore, the chairman and I have prepared an amendment which would restyle our bill to make it an amendment to the 1965 Solid Waste Disposal Act. In addition, we have adopted some provisions contained in the Senate bill which were not included in the House measure.

> The Senate's "solid waste" terminology was agreed to be used instead of the House's "discarded materials" terminology.

> The Senate definitions of solid waste and solid waste management have been included.

> Although the basic House mechanism for State plans and financial assistance were retained, State's were given an overall term of 5 years to close all open dumps.

> An additional authorization of $25 million for each of fiscal years 1978 and 1979 for equipment and construction of rural areas was included.

State grant authorizations were raised by $5 million for each of fiscal years 1978 and 1979. This money can be used for engineering surveys, market studies, and similar projects but it cannot be used for construction or land purchase.

The antitrust exemptions were dropped.

The Justice Department will commence all suits.

Federal facilities will be subject to State law and regulation.

Traditional review procedures were made consistent with the Clean Air and Clean Water Acts.

An interagency committee to conduct a resource conservation study was added.

Authorizations for grants were lowered by $17.5 million, and $35 million was added to authorizations running to the Administrator.

I believe these changes are minor in nature and actually add to the substance and workability of the Solid Waste Act. Therefore, I urge my colleagues to adopt the substitute version of the House bill and to send this measure to the other body for what will undoubtedly be immediate approval and enactment of this legislation into law. 122 *Cong. Rec.* 32598-99 (1976).

Congressman Fred Rooney also stated that:

Mr. Chairman, this amendment is in the nature of a substitute for the text of H.R. 14496.

This substitute text is agreed upon by both the majority and minority of the committee and I have received positive indications from the Senate Public Works Committee that this substitute will be acceptable to them.

As stated the substitute is basically the same as H.R. 14496, with the principal differences being:

First. It's an amendment to the Solid Waste Disposal Act rather than free standing;

Second. It takes the Senate definition of resource conservation.

Third. It adds $25 million a year for grants to rural communities.

Fourth. Permits States to develop their own schedule for closing open dumps within a 5-year period rather than requiring each state to close up its open dumps at the rate of 20 percent per year; and

Fifth. Requires Federal facilities to comply with State and local solid waste plans. *Id.*, at 32631.

In addition, Congressman Brown explained the reasoning which lay behind this House-Senate compromise:

Mr. Chairman, I have had the opportunity to review the substitute which has been drafted and submitted in an effort to expedite the consideration of the slightly different views of the Senate. Considering the very late time in this session this amendment will help us to get Senate concurrence and to get the bill on the President's desk. Mr. Chairman, I am reasonably sure that he will want to sign it. *Id.*, at 32632.

With these explanations, the House passed the House-Senate compromise bill (H.R. 14496) by a resounding vote of 367 to 8. *Id.* Three days later, Senator Randolph introduced the compromise bill in the Senate with the statement that:

The legislation now before the Senate was approved by the House on September 27. It is the result of a merger of my Senate bill and the one reported by the House Committee on Interstate and Foreign Commerce, a bill that was itself drafted as a refinement to S. 2150. I am gratified by the cooperative approach of the House in combining the features of our two bills in this manner. This avoids the need for a conference resolution of differences at this late date in the congressional year.

This comprehensive legislative measure addresses the full range of issues that I perceived when I introduced the Solid Waste Utilization Act in 1975. The measure incorporates the policies of the bill passed by the Senate 3 months ago. Its final approval today by the Senate will advance our country toward more environmentally sound solid waste programs and practices. 122 *Cong. Rec.* 33816 (1976).

Senator Randolph then cataloged the compromises reached between the two chambers in much the same terms as Congressman Rooney had done. *Id.*, at 33817. Following these statements, the Senate adopted the compromise bill without a recorded vote.

On October 22, 1976, President Ford signed the Bill into the law, which has become known as RCRA.

II. THE SOLID WASTE DISPOSAL ACT AMENDMENTS OF 1980

Almost 4 years to the day after RCRA was adopted, on October 21, 1980, President Carter signed into law the Solid Waste Disposal Act Amendments of 1980 (Public Law 96-482,94 Stat. 2334) ("1980 Amendments") which amended RCRA in several important respects: (1) Section 3 confers on the Secretary of Interior exclusive RCRA authority over waste and overburden from surface coal mining; (2) Section 7 exempts from RCRA regulation all oil and gas wastes, coal combustion wastes, cement kiln dust waste, and mining wastes from extraction and processing; (3) Section 8 amends RCRA Section 3002 to require hazardous waste generators to use all reasonable means to assure that the waste arrives at a permitted facility; (4) Section 9 directs EPA to distinguish, "where appropriate", between new and existing facilities in establishing standards for hazardous waste treatment, storage, and disposal; (5) Section 10 clarifies "interim status" so that it may apply to facilities in existence on November 19, 1980; (6) Section 12 expands EPA's RCRA inspection authority; (7) Section 13 expands the scope and penalities available for certain criminal acts; (8) Section 13 also creates a new and unprecedented crime of "knowing endangerment" of human life; (9) Section 17 establishes a new hazardous waste inventory of inactive sites with authority to require monitoring and analysis of such sites which may present a substantial hazard to health or the environment; (10) Section 18 amends RCRA Subtitle D requirements for nonhazardous waste to prohibit "open dumping" effective when EPA has adopted criteria under RCRA Section 1008(a) (3); and (11) Section 25 amends RCRA Section 7003 to give EPA broader authority to take action to abate imminent hazards. These principal amendments are reviewed in the following discussion.* In addition, although none of the 1980 statutory amendments was directed to the question of whether EPA should regulate waste according to their relative "degree of hazard", brief comment on that question is appropriate and is set forth in Subsection (L) of the following discussion.

A. Coal Mining Wastes and Overburden

Many of the 1980 Amendments, including those relating to coal mine wastes, arose as a reaction to EPA's efforts to launch the 1976 RCRA program. As the scope and dimensions of EPA's December 1978 proposed regulations (43 *Fed. Reg.* 58946 [December 18, 1978]) became clear, Congress began reacting to certain aspects of the program which were deemed to be excessive. One of these was the application of RCRA to coal mining wastes already regulated by the Interior Department. The legislative history of the 1980 Amendments reveals widespread support for exempting surface coal mines from the additional burden of RCRA. Congressman Staggers introduced the House version of a coal waste exemption on February 20, 1980. The Stagger's Amendment would treat permits issued under the Surface Mining Control and Reclamation Act of 1977 (SMCRA) as if they were RCRA permits:

---

*   The 1980 Amendments made several other significant changes in RCRA, which space will not permit this chapter to address. These include, among others: clarification (Section 2) that "open dumps" do not include RCRA Subtitle C hazardous waste facilities; creation (Section 4) of a new interagency RCRA coordinating committee; allowance (Section 14) of more stringent state requirements; requiring (Section 24) EPA to share hazardous waste information with OSHA; and imposition of new public participation requirements (Section 26).

Mr. Chairman, this amendment simply provides that the Environmental Protection Agency will defer to the Office of Surface Mining in the Department of the Interior with respect to permits for coal mining wastes and overburden.

The EPA and the Department of the Interior have comparable regulatory authority over hazardous coal mining waste disposal. Final performance standards have already been promulgated by the Office of Surface Mining in the Interior Department of the Surface Mining Control and Reclamation Act of 1977. EPA is required to promulgate hazardous waste regulations under subtitle C of the Resource Conservation and Recovery Act. Final regulations under that statute have not yet been issued.

My amendment provides that all regulatory requirements governing coal mining waste disposal shall be integrated into a single permit to be issued by the Office of Surface Mining. 126 *Cong. Rec.* H. 3358 (1980).

Senators Randolph, Byrd, Huddleston, and Ford co-sponsored an amendment in the Senate to consolidate regulation of coal mining waste by allowing EPA to issue a RCRA "general permit" for all facilities permitted under SMCRA. As Senator Randolph explained it:

It provides that the Environmental Protection Agency will defer to the Office of Surface Mining on permits for coal mining wastes and overburden. Only permits from the Office of Surface Mining will be required. Any requirement which would result from the subtitle (C) hazardous wastes program will be integrated into the Office of Surface Mining permit requirement.

One regulatory program, I think, ought to be sufficient to cope with any hazardous aspects of coal mining wastes. I think it is a realistic approach. It does not attempt to do anything other than give close attention, through the Surface Mining Act, to this problem, which is a very real one. The purpose of authorizing the Administrator to issue a general permit is to allow a single, blanket permit covering all local mining waste sites with permits from the Office of Surface Mining, to substitute for any individual site, hazardous waste permit requirement under the act. 125 *Cong. Rec.* S 13250-51 (1979).

The Conference Committee on the 1980 Amendments followed the House version, but transferred exclusive RCRA authority over SMCRA operators to the Secretary of Interior. Section 3 of the 1980 Amendments provides that:

The Secretary of Interior shall have exclusive responsibility for carrying out any requirement of subtitle C [of RCRA] with respect to coal mining wastes or overburden for which a surface coal mining and reclamation permit is issued or approved under the Surface Mining Control and Reclamation Act of 1977. The Secretary shall, with the concurrence of the Administrator, promulgate such regulations as may be necessary to carry out the purposes of this subsection and shall integrate such regulations with regulations promulgated under the Surface Mining Control and Reclamation Act of 1977.

Section 11 of the 1980 Amendments goes on to provide that the permit covering coal mine waste and overburden issued under SMCRA is to be deemed to be a RCRA hazardous waste permit and that "regulations promulgated by the Administrator under this subtitle shall not be applicable to treatment, storage, or disposal of coal mining wastes and overburden which are covered by such a permit."

In its May 19, 1980 final RCRA regulations, EPA had excluded from the hazardous waste controls "mining overburden returned to the mine site." (§ 260.10(a)(45), 45 *Fed. Reg.* 33075. In addition, although the regulations were silent on the question, EPA's May 19, 1980 Preamble said that "the Agency has deferred regulation of coal mine waste under RCRA." 45 *Fed. Reg.* 33173. The overall effect of these 1980 Amendments is to moot the conditions which EPA would impose for its regulatory exemptions. Instead, coal mine waste and overburden covered by an Interior Department SMCRA permit are now exempt *by statute* from RCRA's hazardous waste program.

B. Mining Waste and Overburden

In addition to exempting coal mine wastes, the 96th Congress widely supported exemptions for oil and gas activities and for all the rest of mining waste and overburden. In May 1979, the Senate Committee on Environment and Public Works reported its bill to amend RCRA which included an exemption for oil and gas wastes for at least 24 months and until Congress adopts legislation approving any EPA-proposed RCRA oil and gas regulation. As the Senate Report stated:

> The Committee determined that the extensive regulatory program proposed by the Agency could have a significant economic impact on domestic oil and gas exploration and production activities. Therefore, regulations on these materials should not be promulgated until further information is developed to determine whether a sufficient degree of hazard exists to warrant additional regulations and whether existing State or Federal programs adequately control such hazards. The Act was amended accordingly. S. Rep. No. 172, 96th Cong., 1st Sess. at 6 (1979).

Also in May 1979, the House Committee on Interstate and Foreign Commerce reported its RCRA amendments which included an absolute and permanent exemption for oil and gas wastes. See H.R. Rep. No. 191, 96th Cong., 1st Sess. at 14 (1979).

Subsequently, on February 20, 1980, Congressman Bevill offered an even broader RCRA amendment, which the House adopted. Mr. Bevill's amendment followed the Senate's 24-month exemption for oil and gas waste and adopted similar exemptions for wastes from coal combustion, cement kiln operations, and ore and mineral mining and processing. See 126 *Cong. Rec.* H 3359-3361 (1980). Mr. Bevill explained the purpose of the amendment:

> The amendment would encourage development of coal as a primary domestic source of energy, avoid unnecessary inflationary impact, and focus the efforts of the Environmental Protection Agency in implementing the Resource and Conservation and Recovery Act toward activities truly necessary to protect public health and the environment; specifically, it would require EPA to defer imposition of regulatory requirements on the disposal of the waste by-product of fossil fuel combustion, of discarded mining materials and of cement kiln dust waste until after EPA has completed studies to determine whether, if at all, these materials present any hazard to human health or the environment. These studies would include evaluation of the economic and environmental aspects of existing and alternative disposal and reuse options. EPA would also be required to focus on the impact of these alternatives on the use of our coal and other natural resources.
>
> Increased use of our Nation's coal supplies as a primary element in our effort to eliminate our reliance on foreign energy sources is an option that we must exercise. We possess the world's largest reserves of coal. We must provide incentives, not disincentives, for its use.
>
> I am not suggesting that increased development of coal resources should occur at the cost of our health or reasonable environmental protection. But I am suggesting that we concern ourselves with removing unnecessary roadblocks to the development of our coal resources.
>
> The effect on coal usage of the regulations EPA has proposed under RCRA clearly would constitute an unnecessary and ill-timed regulatory burden. *Id.*, at H 3361.

With respect to oil and gas waste, the Conference Committee generally adopted the Senate Bill but clarified its intended scope:

> The term "other wastes associated" is specifically included to designate waste materials intrinsically derived from the primary field operations associated with the exploration, development, or production of crude oil, natural gas, or geothermal energy. It would cover such substances as: hydrocarbon bearing soil in and around the related facilities; drill cuttings; materials (such as hydrocarbon, water, sand, and emulsion) produced from a well in conjunction with crude oil, natural gas, or geothermal energy; and the accumulated material (such as hydrocarbon, water, sand, and emulsion) from production separators, fluid treating vessels, storage vessels, and production impoundments.

> The phrase "intrinsically derived from the primary field operations..." is intended to differentiate exploration, development, and production operations from transportation (from the point of custody transfer or of production separation and dehydration) and manufacturing operations.

H.R. Rep. No. 1444, 96th Cong., 2d Sess. at 32 (1980) (hereinafter referred to as "1980 Amendments Conference Report").

With respect to fly ash, cement kiln dust, and mining wastes, the Conference Committee adopted the Bevill amendment with the modification that as to uranium mining wastes, the exemption is limited to uranium mining overburden. *Id.* As a result, Section 11 of the 1980 Amendments expressly provides that "[s]olid waste from the extraction, beneficiation, and processing of ores and minerals, including phosphate rock and overburden from the mining of uranium ore" are excluded from regulation under RCRA until at least 6 months after EPA completes its Section 8002(f) mining waste study and determines whether to promulgate mining waste regulations. The 1980 Amendments also require similar deferral of RCRA regulation pending completion of an EPA study for fly ash and cement kiln dust.

RCRA hazardous waste regulation of drilling fluids, produced waters and other wastes associated with production of crude oil, natural gas, or geothermal energy is deferred for at least 2 years and until after EPA adopts final regulations calling for control of such wastes. In the case of oil, gas, and geothermal wastes, the statute bars EPA from adopting RCRA regulations unless authorized by Act of Congress.

One important consequence of these new statutory exemptions was to moot EPA's May 19, 1980 regulations exempting: (1) overburden for surface mine reclamation (§ 261.4(a)(5), 45 *Fed. Reg.* 33120); and (2) "materials subjected to *in situ* mining techniques which are not removed from the ground as part of the extraction process" (§ 261.4(a)(5), 45 *Fed. Reg.* 33120). In November 1980, EPA adopted a regulation broadly implementing the mining waste exemption. See 45 *Fed. Reg.* 76618 (1980). However, EPA's regulatory Preamble stated that certain wastes generated in connection with mining activities would not be deemed to be exempt and the Agency solicited comments on other aspects of the scope of the statutory exemption. The Preamble states that:

> This [statutory mining waste] exclusion does not, however, apply to solid wastes, such as spent solvents, pesticide wastes, and discarded commercial chemical products, that are not uniquely associated with these mining and allied processing operations, or cement kiln operations. Therefore, should either industry generate any of these non-indigenous wastes and the waste is identified or listed as hazardous under Part 261 of the regulations, the waste is hazardous and must be managed in conformance with the Subtitle C regulations. 45 *Fed. Reg.* 76619 (1980).

As to the overall scope of the statutory exemption, EPA noted that:

> In particular EPA questions whether Congress intended to exclude (1) wastes generated in the smelting, refining and other processing of ores and minerals that are further removed from mining and beneficiation of such ores and minerals, (2) wastes generated during exploration for mineral deposits, and (3) wastewater treatment and air emission control sludges generated by the mining and mineral processing industry. EPA specifically seeks comment on whether such wastes should be part of the exclusion. EPA also seeks comment on how it might distinguish between excluded and nonexcluded solid wastes. *Id.*

C. Distinction Between New and Existing Facilities

When EPA proposed its first RCRA regulations in December 1978, it sought to treat new and existing waste facilities alike (see § 250.41(b)(28), 43 *Fed. Reg.* 58997) and to regulate all existing NPDES wastewater facilities under RCRA (§ 250.45-3, 4, 6, 43

Fed. Reg. 59011-3, 59014-5). The regulated community objected strenuously to this proposal. It was argued that Congress could not have intended that recently completed NPDES facilities, designed and constructed at very substantial cost in order to comply with the 1977 Federal Water Pollution Control Act requirements, be subjected to new, different standards under RCRA. Industry urged that such RCRA standards would require the abandonment of many new NPDES facilities because it would be impossible or impractical to retrofit them. Moreover, EPA's proposed site restrictions (§ 250.43-1, 43 *Fed. Reg.* 59000 (1978) could categorically foreclose continued operation of many existing NPDES facilities. Furthermore, retrofitting other existing hazardous waste facilities would be quite costly and disruptive of manufacturing operations.

Congress responded sympathetically to these concerns about existing facilities. The House Committee on Interstate and Foreign Commerce reported a bill which would have amended the definition of "solid waste" to exclude "solid or dissolved materials in wastewaters received by or discharged from industrial or municipal wastewater treatment facilities..." H.R. Rep. No. 191, 96th Cong., 1st Sess. at 12 (1979).

Subsequently, the sponsor of this original provision, Congressman Swift, offered another amendment on the floor to exempt existing NPDES wastewater facilities when they demonstrate that they do not release significant amounts of hazardous waste to the ground-water. See 126 *Cong. Rec.* H 3354 (1980).

Meanwhile, in May 1979, the Senate Committee on Environment and Public Works reported its bill which included an amendment to RCRA Section 3004 "to authorize the Administration to distinguish in establishing performance standards for hazardous waste treatment, storage and disposal facilities between requirements for new facilities and those for facilities in existence when regulations under this section are promulgated." S. Rep. No. 172, 96th Cong., 1st Sess. at 3 (1979). A colloquy between Senators Randolph and Bentsen elaborated on the Senate's intent:

> Mr. Bentsen: As I said during the committee's discussion of this problem, it would make Congress look a bit ridiculous to pass one law, require the expenditure of a great amount of money, and then pass another law requiring the same expenditure again. It was the committee's intent that the Environmental Protection Agency give special consideration to those facilities where construction or modification was initiated or completed prior to the promulgation of regulations associated with this bill. The committee would anticipate that the Environmental Protection Agency would require a far lower degree of control for these existing facilities than for new facilities.
> Mr. Randolph: The Senator is correct in his assessment of the committee action and intent. 125 *Cong. Rec.* at S 13243-44 (1979).

The Conference Committee was then faced with somewhat different approaches to this issue: The House bill would have statutorily exempted existing facilities from RCRA Section 3004 standards when they satisfied a ground-water contamination test. The Senate version would have simply authorized EPA to distinguish between new and existing facilities. The Conference adopted a compromise position:

> The conference substitute amends Section 3004 to provide that the Administrator should, where appropriate, establish separate requirements for new and existing facilities...
> In including this amendment in the conference bill, the conferees are concerned primarily with two key factors which distinguish existing facilities from new facilities. First, existing facilities are already committed to particular sites, and thus relocation especially where the facility is directly connected to a manufacturing operation (like a wastewater treatment impoundment), is usually not a feasible alternative. Second, existing facilities have made substantial commitments to the design and construction of the facility.
> The conferees recognize that there are many practical and technical limitations in modifying or retrofitting existing facilities to meet standards which may be reasonable to apply to new

facilities. If applicable performance standards can be met with the existing design, it seems unduly burdensome to require some modifications or retrofitting. This may be particularly true for industrial wastewater treatment facilities where the owner would not only be faced with the tremendous cost associated with reconstructing his existing ponds, or building replacement ponds; but in many cases would find that his manufacturing plant would have to close while his impoundment was being retrofitted, since the impoundment is needed to treat his wastewater to satisfy his Clean Water Act discharge permit.

The Adminstrator should exercise discretion with regard to the standards set forth for existing as differentiated from new facilities, including establishment of separate requirements for new and existing facilities where necessary taking into consideration the factors noted above. Regulations issued under Section 3004 should provide sufficient flexibility to allow differing designs and locations for existing facilities as long as the health and environmental performance objectives can be met.

The conferees believe that distinctions between new and existing facilities, can, and should, be made without compromising on the level of protection necessary to protect public health and the environment. 1980 Amendments Conference Report at 33-34.

D. Interim Status

RCRA Section 3005 allows facilities which are existing and which gave the required preliminary notification on August 18, 1980 to qualify for "interim status" and thus continue operating while a RCRA permit application is pending. Since it may take 5 years or more before a permit application is acted upon, interim status has obvious importance. Section 10 of the 1980 Amendments changes the date for determining whether a hazardous waste facility is "existing" for purposes of interim status. Prior to the 1980 Amendments, only facilities in existence on October 21, 1976 could qualify. Now all facilities in existence as of the effective date of the RCRA regulations — November 19, 1980 — can qualify for interim status.

The Conference Report accompanying the 1980 Amendments explains further that:

> The conferees intend that a facility need not actually be in operation and receiving wastes to be "in existence". It must, however, have obtained all necessary State, local or Federal permits and clearances, and, being justified in relying on those permits, the owner or operator must have a made a financial commitment which cannot be terminated, relocated, or modified without a substantial loss. The meaning of the phrase "in existence" can best be understood by reviewing the discussion of the definition of "commenced construction" in the report of the Committee on Environmental and Public Works on the Clean Air Act Amendments of 1977 (Senate Report Number 95-127). 1980 Amendments Conference Report at 34.

EPA has adopted regulations interpreting and applying these amendments. See 46 *Fed. Reg.* 2344 (1981).

E. RCRA Inspection Authority

As adopted in 1976, RCRA Section 3007 provided EPA officers and employers with authority to obtain access to documents relating to hazardous wastes and to sample such wastes "[f]or purposes of developing or assisting in the development of any regulation or enforcing the provisions of this subtitle." This provision did not authorize EPA's independent contractors to exercise EPA's inspection authority; nor did it authorize inspections for purposes other than those related to Subtitle C; and it did not apply to persons who had in the past, but no longer, operated hazardous waste facilities. In order to overcome these deficiencies in the existing law, both the House and Senate adopted amendments which would expand Section 3007 to cover contractors and include inspections for certain purposes beyond Subtitle C. See H.R. Rep. No. 191 96th Cong., 1st Sess. at 15-16 (1979); S. Rep. No. 172, 96th Cong., 1st Sess. at 14 (1979).

The Conference Committee generally followed the Senate bill and included a provision making it a crime for any person not covered by the Trade Secrets Act (18 U.S.C.

§ 1905), such as EPA's contractors, to disclose confidential information obtained under Section 3007. The new amendment also requires that, notwithstanding any confidentiality claim, Section 3007 information must be made available on request to any Congressional committee.

F. Additional Criminal Penalties

The 1980 Amendments also expand RCRA's criminal sanctions.

Since enactment in 1976, RCRA has provided for civil penalties of up to $25,000 per day for violations of its requirements (RCRA § 3008 (a)), for revocation or suspension of RCRA permits for noncompliance (RCRA § 3008 (b)) and for criminal fines and/ or imprisonment for: (1) transporting hazardous waste to a nonpermitted facility; (2) treating, storing, or disposing of hazardous waste without a RCRA permit; and (3) making false statements in RCRA records and reports (RCRA § 3008 (d)).

The 1980 Amendments increase the penalties and expand the scope of criminal sanctions in several significant ways, as described below.

1. Increased Criminal Penalties for Permit Violations

Existing law provided for misdemeanor penalties (up to 1 year in prison and a $25,000 fine) for all first time criminal violations of RCRA. The 1980 Amendments double these penalties for violations of RCRA § 3008(d)(1) (transport to a nonpermitted facility) and § 3008(d)(2) (failure to have or comply with a RCRA permit).

2. New Criminal Sanctions for RCRA Permit Violations

Criminal sanctions now expressly apply to treating, storing, or disposing of hazardous waste "in knowing violation of any material condition or requirement of such permit..." This amendment was intended to clarify the existing provision against operation without a permit to make certain that violation of permit terms can also give rise to criminal sanctions. The Conference Report to the 1980 Amendments states that:

> This section is intended to prevent abuses of the permit system by those who obtain and then knowingly disregard them. It is not aimed at punishing minor or technical variations from permit regulations or conditions if the facility operator is acting responsibly. The Department of Justice has exercised its prosecutorial discretion responsibility under similar provisions in other statutes and the conferees assume that, in light of the upgrading of the penalties from misdemeanor to felony, similar care will be used in deciding when a particular permit violation may warrant criminal prosecution under this Act. 1980 Amendments Conference Report at 37.

3. New Criminal Sanctions for Destroying RCRA Records

In addition, the 1980 Amendments create a new criminal provision applicable to anyone who "knowingly generates, stores, treats, transports, disposes of, or otherwise handles any hazardous waste (whether such activity took place before or takes place after the date of the enactment of this paragraph) and who knowingly destroys, alters, or conceals any record required to be maintained under regulations promulgated by the Administrator under this subtitle...."

G. New Crime of "Knowing Endangerment"

Finally, the 1980 Amendments also establish a new crime for any person who knowingly: (1) transports waste to a nonpermitted facility; (2) violates permit terms; (3) fails to disclose material facts in permit applications; or (4) fails to comply with interim status requirements, when the person engaged in these activities acts with inexcusable disregard or extreme indifference for human life, and knows at that time that "he thereby places another person in imminent danger of death or serious bodily injury...." The penalty prescribed for such conduct is a fine of up to $250,000 for individuals, and $1 million for "organizations", and/or imprisonment for up to 2 years (5 years if the conduct manifests "extreme indifference for human life"). The statute elaborates in detail the various elements of this new offense, particularly the requisite *mens rea*. See, RCRA Section 3008(f).

This new criminal provision was drafted by the Department of Justice and introduced in the House at the Department's request by Congresswoman Mikulski. See 126 *Cong. Rec.* H 3367 (1980). Congressman Florio explained the provision in response to questions about its scope:

> Of course, the whole question will be determined by a court, and the feeling is that, on the actual person who violates the standard, which is reckless disregard, who should have known of the inappropriateness of the disposal. Obviously, these are factual matters and we have had instances in the past whereby someone has had a release ostensibly absolving them from any responsibility of inappropriate disposal; under this statute, of course, and a very strict standard of law in the criminal statutes, there will be a need to go beyond just the front of the release to find out whether that individual should have had knowledge as to the accuracy or adequacy of the disposal producer. So in effect what I am suggesting, not a direct answer, a factual determination will have to be made by the law enforcement agencies through the indictment process. 126 *Cong. Rec.* H 3368 (1980).

As originally drafted, this provision applied to "reckless endangerment" of human health. The Senate bill (S. 1156) contained no similar provision and the House-Senate conferees modified the House bill to cover only "knowing endangerment." (1980 Amendments Conference Report at 37-38.) The Conference Report explained the purpose and scope of this new provision as follows:

> The purpose of this new section is to provide enhanced felony penalties for certain life-threatening conduct. At the same time, the new offense is drafted in a way intended to assure to the extent possible that persons are not prosecuted or convicted unjustly for making difficult business judgments where such judgments are made without the necessary scienter.

> The knowledge necessary for culpability of a natural person is actual knowledge, which may be established by direct or circumstantial evidence, but not constructive or vicarious knowledge.

> In either event, the endangerment offense depends upon a showing that a natural person actually knew that his conduct at that time placed another person in imminent danger of death or serious bodily injury.

> For the conduct described above to constitute the crime of endangerment, the government must also prove that the defendant's conduct met one of the two "tiers" of culpability defined in subparagraphs (e) (2) (A) and (B). The more egregious conduct — that manifesting an "extreme indifference" to human life — is punishable by up to 5 years imprisonment or a fine of $250,000 (one million dollars for an organizational defendant), or both. An obvious example of such conduct would be the dumping of what is known to be poisonous waste, in potentially lethal amounts, into what the defendant knows is a municipal drinking water supply, even if, due to a mere quirk of fate, severe bodily injury or death does not in fact result.

> The other test for culpability, for which a somewhat lesser but still severe penalty is set, is conduct manifesting an unjustified and inexcusable disregard for human life. This violation carries a penalty of $250,000 (again, one million dollars for an organizational defendant) or 2 years imprisonment, or both. Such conduct must include a conscious disregard for human life that is neither excusable nor justified by countervailing considerations.

> It is not the purpose of this amendment either to create criminal liability or to impose enhanced penalties for errors in judgment made without the necessary scienter, however dire may be the danger in fact created.

> There is also general recognition that serious criminal charges are not an appropriate vehicle for second-guessing the wisdom of judgments that are made on the basis of what was known at the time where the person acted without the necessary element of scienter.

. . . The conferees believed that the responsibility for the felony of criminal endangerment should properly be confined only to those persons who themselves have actual knowledge of the danger resulting from their conduct. A supervisor, for example, who personally lacks the necessary knowledge, should not be criminally prosecuted for knowledge that only his subordinates possessed. Thus, whether or not vicarious knowledge may be sufficient for other crimes, the new endangerment offense under subsection (e) applies only to those people who have personal knowledge of the danger their conduct created.

. . . This provision does not deal with the separate question of establishing corporate liability. The criminal responsibility of a corporation for knowledge possessed by its officers and managing agents should be governed by traditional principles. (1980 Amendments Conference Report at 37-40.)

Notwithstanding the Conference Committee's clarifications, in view of the complexity and novelty of this new crime, it is very likely to be the subject of judicial challenges and review. Even a cursory reading of this criminal statute brings to mind the Supreme Courts oft-repeated maxim that:

It is settled that, as a matter of due process, a criminal statute that "fails to give a person of ordinary intelligence fair notice that his contemplated conduct is forbidden by the statute," *United States v. Harris,* 347 U.S. 612, 617 (1954), or is so indefinite that "it encourages arbitrary and erratic arrests and convictions," *Papachristou v. City of Jacksonville,* 405 U.S. 156, 162, 92 S.Ct. 839, 843, 31 L.Ed.2d 110 (1972), is void for vagueness.

See generally *Grayned v. City of Rockford,* 408 U.S. 104, 108-109, 92 S.Ct. 2294, 2298-2299, 33 L.Ed. 2d 222 (1972); *Colautti v. Franklin,* 439 U.S. 379 (1979).

H. New Hazardous Waste Inventory

As originally adopted, RCRA was prospective only and did not cover inactive hazardous waste sites. To address this problem, in May 1979, the House Committee on Interstate and Foreign Commerce reported its bill with a new provision for mandatory state-conducted inventories of all sites where hazardous wastes have at any time been stored or disposed of (RCRA Section 3012) and for EPA-ordered monitoring and analysis of certain such sites (RCRA Section 3013). As the House Report stated:

A new issue has arisen which was not evident in 1976: the problem of abandoned hazardous wastes disposal sites. This discovery led to an increased awareness of the gaps in RCRA under Subtitle C, which primarily addressed the cradle-to-grave management of hazardous wastes in new and existing disposal sites. In an attempt to narrow this gap, the Committee amended RCRA to include a new state-wide inventory program (Section 3012) for abandoned hazardous wastes disposal sites and to clarify that the Administrator has authority to take action with regard to abandoned and inactive sites.

This measure is a direct result of concerns expressed during reauthorization hearings. There is agreement that some preliminary measures are needed to immediately address the abandoned sites issue. Some determination of the scope of the problem is required before a new or expanded program can be launched. This provision should be viewed as an initial step toward addressing the abandoned sites problem, and not as a solution. H.R. Rep. No. 191, 96th Cong., 1st Sess. at 4 (1979).

Section 3(k) amends Subtitle C to add a new Section 3012 which directs the states to undertake an inventory of hazardous sites as expeditiously as practicable. The inventory is to include a description of the site and as much information as is practicably obtainable regarding the nature and the extent of the health hazard it presents. If a state fails to comply adequately with these provisions, the Environmental Protection Agency is empowered to carry out the inventory program in that state. A grant program is established, with an authorization level of $20,000,000 to make funds available to the states to carry out this program. States which have already conducted such an inventory may apply for grants for reimbursement for all or a portion of the costs incurred.

A new Section 3013 amends Subtitle C to require that certain specific information be included in the state inventory. The Administrator is required to make available to the states such information as he has available concerning the specified items required to be included in the inventory. Upon the receipt of any information indicating that a hazardous waste facility on a site may create a significant hazard to human health or the environment, the Administrator may issue an order requiring the persons who owned or operated the facility or site for any period during which hazardous waste was treated or disposed of to perform necessary monitoring, testing, analysis, and reporting or to pay for the costs of these procedures. *Id.,* at 10.

The bill adopted by the House Committee contained several features with respect to monitoring which were very troublesome to the regulated community. This original version provided that:

> Upon the receipt of any information indicating that hazardous waste is, or has been, stored, treated, or disposed of at any facility or site, and the presence of any hazardous waste at such facility or site, or the release of any such waste or other substance from such facility or site, may create a significant hazard to human health or the environment, the Administrator may issue an order requiring the persons who owned or operated such facility or site for any period during which hazardous waste was treated or disposed of at such site to —
> (1) conduct such monitoring, testing, analysis, and reporting as the Administrator deems necessary to ascertain the nature and extent of the potential hazard to public health and the environment associated with such facility or site; or
> (2) pay for the costs of such monitoring, testing, and analysis.... *Id.,* at 19.

The regulated community was seriously concerned by three aspects of this bill: (1) the "trigger" for monitoring was quite imprecise and did not require that EPA make any determination of hazard; (2) persons who operated a site for any period could be forced to pay the entire monitoring cost for the site; and (3) the person obligated to monitor was given no opportunity to develop his own monitoring program.

Several months after the House Report was issued, and following extended negotiations among House staff, EPA, and the Chemical Manufacturers Association, the sponsor of the original amendment, Congressman Gore, offered a further change in the proposed new RCRA Section 3013. 126 *Cong. Rec.* at H 3356-7 (1980). Mr. Gore's revised proposal differed from the original version in significant respects: (1) it changed the standard by which monitoring would be triggered from "may create a significant hazard" to "may present a substantial hazard"; (2) it required that monitoring be triggered only when "the Administrator determines" that the situation may present such a hazard; (3) it changed the method for allocating the cost of monitoring from allocation of the entire cost to any previous owner-operator, to allocation of the entire cost to the most recent prior owner with knowledge of the site when the existing owner-operator "could not reasonably be expected to have actual knowledge of the presence of hazardous waste;" and (4) instead of having EPA order such monitoring as it saw fit, the amendment allowed the person liable for monitoring costs to propose a monitoring system, subject to EPA approval. The Conference Committee adopted the modified House provision and added the direction that remedial actions not be postponed pending the hazardous waste inventory.

On December 11, 1980, the so-called "Superfund" statute (Public Law 96-510, 94 Stat. 2767) was signed by President Carter. The Superfund Act also provides for a hazardous waste inventory, by requiring that by June 9, 1981:

> [A]ny person who owns or operates or who at the time of disposal owned or operated, or who accepted hazardous substances for transport and selected, a facility at which hazardous substances (as defined in section 101(14) (C) of this title) are or have been stored, treated, or disposed of *shall,* unless such facility has a permit issued under, or has been accorded interim status under, subtitle C of the Solid Waste Disposal Act, *notify the Administrator of the Environmental Protection Agency of the existence of such facility,* specifying the amount and type of any hazardous substance to be found there, and any known, suspected, or likely releases of such substances from such facility. Pub. L. 510, § 103(c).

Failure to so notify EPA is punishable by fine up to $10,000 or imprisonment for up to 1 year. The mandatory reporting program called for by Superfund overlaps with the mandatory inventory which RCRA Section 3012 directs the states to conduct.

I. Open Dumping Prohibited When Criteria Adopted

Subtitle D of RCRA provides for a regulatory system applicable to nonhazardous solid wastes. The basic structure turns upon a prohibition of "open dumping" which EPA is authorized to define by regulation. Under the 1976 version of Subtitle D, Section 4005, there remained some uncertainty as to when the open dumping prohibition became effective. Some commentators had contended it could only become effective after EPA had completed and published its dumpsite inventory under Section 4005(b). The Agency contended that the prohibition took effect when EPA promulgated the "criteria" under Section 1008(a)(3) for determining what constitutes "open dumping". At the Agency's behest, both the House and Senate Committees considering RCRA amendments, and subsequently the Conference Committee, included an amendment to Section 4005 to clear up the uncertainty and make the "open dumping" prohibition effective when the Section 1008 "criteria" have been promulgated. Since final Section 1008 criteria were published in September 1979 (44 *Fed. Reg.* 53438), the practical consequence of this amendment is to prohibit "open dumping" as of September 13, 1979.

J. Expansion of RCRA's Imminent Hazard Authority

Under the 1976 version of RCRA Section 7003, EPA was authorized to bring an action in Federal Court for an injunction to stop hazardous waste handling, transportation, treatment, storage, or disposal which "is presenting an imminent and substantial endangerment to health or the environment."

EPA was apparently concerned that this authority was insufficient because the standard for action — "is presenting" — was too stringent and because EPA was required to go to court in every such case — it had no authority to issue Administrative orders requiring remedial action. Persuaded that these were significant deficiencies, the House Committee included amendments to Section 7003 to accomplish both objectives. See H.R. Rep. No. 191, 96th Cong., 1st Sess. at 25-26 (1979). The House Report explained these amendments as follows:

> The committee believes that the Administrator has not been sufficiently vigorous in using the authority under section 7003 to act in those circumstances which pose an imminent and substantial endangerment to health or the environment. In hearings before the committee the agency acknowledged that the authority had not been used fully, but that the agency has recently filed several legal actions under this section and intends to expand its enforcement program. The committee endorses this intention and the reported bill adds language to the Act to allow the Administrator to act upon receipt of evidence that the handling, storage, treatment, transportation, or disposal of solid waste or hazardous waste may present such an imminent and substantial danger. The committee intends that the Administrator use this authority where the risk of serious harm is present. The committee heard numerous witnesses testify to the dangers and risks associated with hazardous sites which are now abandoned and inactive, as well as active sites. The Administrator's authority under section 7003 to act in situations presenting an imminent hazard should be used for abandoned sites as well as active ones.
>
> Because the committee believes that the Administrator should have adequate enforcement tools for carrying out the purposes of section 7003, the reported bill includes a provision which gives authority to the Administrator to issues such orders as may be necessary to protect the public health and the environment under circumstances which may present an imminent and substantial endangerment. *Id.,* at 4-5.
>
> *Section 3(p)* amends section 7003 to allow the Administrator to take action upon receipt of evidence that the handling, storage, treatment, transportation, or disposal of any solid or

hazardous waste may present an imminent and substantial endangerment to health or the environment. The Administrator is also given emergency powers, to issue such orders as may be necessary to protect the public health and a penalty is included for one who willfully violates an order of the Administrator. *Id.,* at 11.

The Senate Committee bill did not make comparable changes in Section 7003 and the Conference Committee adopted the House version. Accordingly, the standard for action under Section 7003 has been changed from "is presenting an imminent and substantial endangerment to health or the environment" to "may present" such endangerment. In addition, the 1980 Amendments authorize EPA to remedy such situations either by bringing an action for an injunction in District Court or by issuing an Administrative order. Under new Section 7003(b), such EPA orders are now subject to independent enforcement by EPA by fine of not more than $5000/day.

K. Degree of Hazard

No discussion of the 1980 Amendments would be complete without reference to the so-called "degree of hazard" issue. Although the 1980 Amendments do not directly address this question, the degree of hazard issue crept into the floor debate and interested parties will no doubt call upon these discussions to support their positions.

The issue first arose in December 1978, when EPA proposed comprehensive RCRA Subtitle C regulations. These proposals would have established uniform reporting, transport, treatment, storage, and disposal standards applicable to all "hazardous wastes" without regard to the degree of hazard posed by each such waste. The regulated community argued that both the statute and its legislative history required that EPA take into account the degree of hazard posed in setting the applicable standards.

1. Industry Argued That RCRA Expressly Requires Consolidation of Degree and Duration of Hazard

RCRA Section 3004 provides for standards for the treatment, storage, or disposal of hazardous wastes *"as may be necessary* to protect human health and the environment" (emphasis added). Such standards thus appear to contemplate that the degree of hazard posed by a waste be considered in order to satisfy the statute's directive that standards be promulgated *"as may be necessary".* Uniform standards applicable to all hazardous wastes apparently would not satisfy this directive.

Moreover, Section 3004 standards are required to include standards for "financial responsibility". RCRA Section 3004(6). The final sentence of Section 3004 provides that:

> No private entity shall be precluded by reason of criteria established under paragraph (6) from the ownership or operation of facilities providing hazardous waste treatment, storage, or disposal services where such entity can provide assurances of financial responsibility and continuity of operation consistent with the *degree and duration of risks* associated with the treatment, storage, or disposal of specified hazardous waste. (Emphasis added.)

EPA is expressly required to accept assurances of financial responsibility which are "consistent with the *degree and duration of risks* associated with the treatment, storage, or disposal *of specified hazardous waste.*" In other words, the required financial responsibility must be related to the degree and duration of risks of the particular waste. It seems impossible for EPA to comply with this statutory directive without taking into account "degree and duration of risks" in setting treatment, storage, and disposal standards, and the concomitant level of financial responsibility.

2. Industry Also Argued That the Legislative History Requires Consideration of Degree of Hazard

EPA has frequently identified the potential migration of leachate as one of the principal hazards addressed by RCRA (see Preamble, 43 *Fed. Reg.* at 58952, col. 2 (1978); Resource Conservation And Recovery Act of 1976: Hearings on H.R. 14496 Before the Subcomm. on Transportation And Commerce of The House Comm. on Interstate And Foreign Commerce, 94th Cong., 2d Sess. 98 (1976) [hereinafter "House Commerce Hearings"].

In describing the potential for hazardous wastes to leach, EPA Deputy Assistant Administrator for the Office of Solid Waste Management Programs, Sheldon Meyers, testified that leaching was dependent on "waste type". As Mr. Meyers stated:

> The transport, attenuation, and impact of pollutants such as leachate, in the subsurface are complex issues. In general, they appear to be related to climate, soil characteristics, disposal site operational characteristics, *waste types,* site hydrogeology, and water use in the area. House Commerce Hearings at 99. (Emphasis added.)

Thus, one of the principal risks associated with hazardous wastes on land is itself dependent on several factors, including the type of waste. Apparently, different standards to control leaching were contemplated according to the different "waste types."

In addition, in his testimony before a Senate Committee, Deputy Assistant Administrator for the Office of Solid Waste Management Programs, Arsen Darnay, testified that the standards for hazardous waste treatment, storage, and disposal sites would be established on a case-by-case basis, one chemical at a time:

> Senator McClure. You talk about the proposal of Federal licensure of hazardous waste disposal. Is there any minimum quantity of relative test concerning the public danger of public hazard before the Federal licensure would come as a requirement?
>
> What I am thinking of is the relatively casual, relatively dispersed, relative lack of concentrated disposal of hazardous waste.
>
> Mr. Darnay. We would like to see established, Senator McClure, sites that are closely controlled by the State or by the Federal Government, *depending upon what kind of waste it is.*
>
> Senator McClure. Well, with one exception and that is when you say that the scientific tests indicate that they are potentially hazardous; is there a quantity test to determine that or are you looking simply at the character of the waste being disposed of, saying that one ounce is equally dangerous as a ton?
>
> Mr. Darnay. Under our proposed legislation, *we would be required to make a case-by-case determination which would include the designation of levels of concentrations.* Obviously, it would vary from chemical to chemical. In some cases you would know that the chemical has been biologically accumulated and would therefore show up in some food substance, like mercury. You probably recall the mercury in swordfish — mercury does accumulate biologically.
>
> There we might not pay so much attention to concentration where dilution is an approach. *We would establish these standards one at a time on one chemical at a time or one waste stream at a time based on tests as to what is an acceptable or unacceptable level.*

The Need For a National Materials Policy: Hearings Before The Panel on Materials Policy of The Subcomm. on Environmental Pollution of the Senate Comm. on Public Works, 93d Cong., 2d Sess. 87 (Part 1, 1974) (hereinafter referred to as "National Materials Policy Hearings") (Emphasis added.)

Many commentators also proposed that a degree of hazard system be included in the final regulations in order to: (1) prioritize our response to the hazardous waste problem by first addressing those wastes which pose the greatest risks and thereby reducing the anticipated shortage of suitable waste treatment facilities; (2) tailor the required waste management standards to fit more closely the particular hazard presented; and (3) set small quantity exemptions which took into account the degree of hazard posed. See generally 45 *Fed. Reg.* 33164 (1980).

EPA rejected these arguments and refused to adopt a degree of hazard system because:

> (1) The Agency does not believe that any of the degree of hazard systems suggested by commentors [or any of the Agency could itself conceive] are capable of actually distinguishing different degrees of hazard among the myriad hazardous wastes and also reasonably relating management standards to these degrees in a technically and legally defensible way.
> (2) The Agency believes that the final regulations already achieve the objectives of a degree of hazard system; thus, such a potentially complex and challengable system is unnecessary. *Id.*

On June 4, 1979, in Senate floor debate on the 1980 Amendments, Senator Randolph, Chairman of the Senate Committee on Environment and Public Works, which had reported a bill to amend RCRA, remarked that:

> One issue not directly addressed in these amendments has come to my attention. In the public comments on the proposed regulations under subtitle C, many commentators suggested schemes for the classification of hazardous waste, under section 3001 according to the degree of hazard. Such an approach is not permissible or appropriate under section 3001, nor does the reported bill contain any amendment which would provide that flexibility. It would be unfortunate if the Environmental Protection Agency were to repropose any of the subtitle C regulations based on this "degree of hazard" theory. Much time would be lost in an area that urgently needs regulation, and the approach would be unlikely to withstand judicial challenge.
> On the other hand, the performance standards to be established under section 3004 should contain whatever is necessary to protect health and the environment from the hazardous waste in question, in any particular disposal method or circumstance. This is the appropriate point at which the regulations can reflect the degree of hazard of a waste or a disposal practice. That flexibility can be accomplished within the current agency consideration of the proposed regulations and the comments received, in a manner consistent with the statute. 125 *Cong. Rec.* S. 13242 (1979).

First, it is unclear what effect Senator Randolph's remarks would have on a proposed degree of hazard approach. The major purpose of such an approach was to tailor management requirements to the particular hazard posed. For example, if waste A were extremely toxic and highly mobile in ground-water, and waste B were not, there would seem to be sound reasons for managing waste A more stringently than waste B. Apparently Senator Randolph's remark would not preclude such an approach.

Second, however, the 1980 Amendments did not address the degree of hazard question and the legislative history of the 1976 statute does not support Senator Randolph's view that waste classification by degree of hazard is unauthorized by Section 3001. Accordingly, to the extent they would limit EPA's authority, Senator Randolph's remarks in June 1979, during the 96th Congress, would seem to have little weight in determining the intent of the 94th Congress when it adopted RCRA in 1976. As the Supreme Court has recently stated in *Consumer Product Safety Commission* v. *GTE Sylvania, Inc.,* 447 U.S. 102, 117 (1980): "In evaluating the weight to be attached to subsequent legislative statements, we begin with the oft-repeated warning that 'the views of a subsequent Congress form a hazardous basis for inferring the intent of an earlier one.' (Citations omitted). Although it has been said that a subsequent statute's declaration of the earlier statute's intent is entitled to weight:

> The less formal types of subsequent legislative history provide an extremely hazardous basis for inferring the meaning of a Congressional enactment. While such history is sometimes considered relevant, this is because, as Chief Justice Marshall stated in *United States* v. *Fisher,* 2 Cranch 358, 386, 2 L.Ed. 304 (1805): "Where the mind labours to discover the design of the

legislature, it seizes everything from which aid can be derived.'' *See Andrus v. Shell Oil Co;* - 446 U.S. 657, 666, n.8 (1980). Such history does not bear strong indicia of reliability, however, because, as time passes, memories fade and a person's perception of his earlier intention may change. Thus, even when it would otherwise be useful, subsequent legislative history will rarely override a reasonable interpretation of a statute that can be gleaned from its language and legislative history prior to its enactment. 447 U.S. at 118 n.13.

Again, in its recent decision in *United States v. Clark,* 445 U.S. 23, 33, n.9 (1980), the Supreme Court held:

> In any event, the views of some Congressmen as to the construction of a statute adopted years before by another Congress have "very little, if any significance" *United States v. Southwestern Cable Co.,* 392 U.S. 157, 170 (1968) (quoting *Rainwater v. United States,* 356 U.S. 590 (1958).

Third, even if Senator Randolph's remark had been contemporaneous with the adoption of RCRA in 1976, no similar remark is contained in the report of the Senate Committee that Senator Randolph chaired:

> And ordinarily even the contemporaneous remarks of a single legislator who sponsors a bill are not controlling in analyzing legislative history.

Chrysler Corp. v. Brown, 441 U.S. 281, 311, 99 S.Ct. 1705, 1722, 60 L.Ed. 2d 208 (1979). *Consumer Product Safety Commission v. GTE Sylvania, Inc.,* 447 U.S. 102 at 118.

The reasons for this rule are quite sound: The personal views of members which are not incorporated into deliberate documents like Committee reports simply do not reflect a consensus; they are not the result of the legislative process. See e.g., *United States v. Gila River Prima-Maricopa Indian Community,* 586 F. 2d 209, 215 (Ct. Cl. 1978). Thus, whatever RCRA may require or permit with respect to a degree of hazard approach, Senator Randolph's June 1979 remarks should have little legal significance.

III. HAZARDOUS AND SOLID WASTE AMENDMENTS OF 1984

A. The Unique Legislative History of the 1984 Amendments

The 1984 amendments to RCRA arose out of an unprecedented legislative and political process. In 1980, Ronald Reagan was elected President of the U. S. on a platform calling for the deregulation of American industry. In particular, the President had expressed the strong sentiment that environmental regulation had gone too far. The President's initial cabinet and EPA appointments seemed to carry out this theme. Ann Gorsuch was appointed Administrator of EPA and James Watt was appointed Secretary of the Department of Interior. Ms. Gorsuch was a Republican member of the Colorado legislature without any significant background in environmental control. Ms. Gorsuch appointed Rita Lavelle as Assistant Administrator of EPA in charge of the hazardous waste program.

In the early days of the first Reagan Administration, there was widespread controversy associated with the environmental and natural resources policy statements made by Ms. Gorsuch and Mr. Watt. In addition, EPA failed to carry forward on the initial regulatory steps launched by the Carter administration under RCRA and a variety of regulatory deadlines passed unfulfilled.

Superimposed over this controversy, in 1982 and 1983 the public and Congress experienced a severe crisis of confidence in EPA's leadership. Serious questions arose about EPA's enforcement of RCRA and Superfund. To begin with, Congress investigated EPA's administration of the new Superfund law and issued a subpoena requiring the production of EPA documents with respect to 160 hazardous waste sites scheduled for early clean-up under that law. In response to that subpoena, and upon orders from President Reagan, EPA Administrator Gorsuch withheld 74 documents under a claim of executive privilege. By withholding the documents sought by the House Committee, the President and EPA Administrator Gorsuch came into direct conflict with the House of Representatives. In response, on December 16, 1982, the House of Representatives voted, 259 to 105, to hold Gorsuch in contempt of Congress for failure to produce the requested documents.

In the following months, the controversy continued to seethe, reaching a crescendo in March 1983. Questions were raised about the propriety of a variety of acts of Assistant Administrator Lavelle. These questions included serious concerns about alleged conflicts of interests. On February 7, 1983, the President finally, and somewhat belatedly, fired Rita Lavelle. The firing of Lavelle did not stem the tide of congressional and popular criticism, however. The controversy continued to increase, and, on March 9, 1983, under severe public and congressional pressure, EPA Administrator Ann Gorsuch (then Ann Burford as a result of her marriage to Bureau of Land Management Director, Robert Burford) was forced to resign. On March 21, 1983, the President named former EPA Administrator Ruckelshaus to resume his prior position.

On May 18, 1983, the House of Representatives voted unanimously (413 to 0) to hold Rita Lavelle in contempt of Congress for failing to appear in response to a subpoena from the House Subcommittee investigating charges of conflict of interest at EPA. Lavelle was indicted on the charge of contempt of Congress, but was acquitted by a jury in Washington, D.C. on July 22, 1983. On August 4, 1983, Lavelle was again indicted on charges of perjury and obstructing a congressional probe into the solid waste program. In particular, Lavelle was accused of lying under oath about when she first knew that her former employer, Aero-Jet General Corp., had been involved in the so-called Stringfellow Acid Pits dump site in California. On December 1, 1983, a federal grand jury in Washington, D.C. found Lavelle guilty on four of the five counts on which she had been indicted. On January 9, 1984, Lavelle was sentenced to 6 months in prison and fined $10,000 in connection with this conviction.

The sum of all of these startling events was to erode congressional and public confidence in the administration of the Environmental Protection Agency to its lowest point in its relatively short history. Congress did not believe that EPA was administering the hazardous waste laws as it intended, and it believed that the administrators of the program had not been truthful in testifying about their activities with respect to those programs. As a result, both the House and Senate went about the job of crafting the 1984 Amendments to RCRA with a tremendous distrust of EPA and a vengeful intent to plug what they perceived as ''loopholes'' in the existing statutory structure which had permitted EPA officials to undermine their 1976 mandate.

Perhaps in no other regulatory program in recent history has the congressional ire been raised so high. As a result, Congress not only adopted unprecedented, detailed regulatory-type statutory standards, but it also imposed extraordinary action-forcing procedures.

Congress was concerned principally that its original intent in enacting RCRA in 1976 had been eroded by a variety of ''loopholes'' in existing law. Congress estimated that approximately half of all hazardous waste generated in the U.S. (about 40 million metric tons/year) was escaping control through these ''loopholes''. First, boilers which burned hazardous waste for purposes of recovering the energy value of the wastes were essentially unregulated under RCRA, even though the emissions from such boilers posed serious health risks and even though EPA already had authority to address those risks. Second. EPA's regulations had excluded from all regulatory control so-called "small-quantity generators" of hazardous waste who generated 1000 kg or less of such wastes per month. The Congressional Office of Technology Assessment had estimated that the small generator exemption alone allowed approximately 4 million metric tons of hazardous waste per year to escape regulatory control.

Third, EPA had failed to promulgate regulations governing waste oil used as a fuel and as a dust retardant. This failure to regulate was estimated to exclude from environmental controls another 4 million metric tons of hazardous wastes per year. Fourth, EPA had failed, in Congress' judgment, adequately to implement its existing authority under RCRA and the Safe Drinking Water Act to regulate the underground injection of hazardous waste. Congress was particularly concerned that with a clamp-down on the disposal of hazardous waste through surface impoundments and other means, underground injection would become more prevalent. In particular, EPA had declined to regulate the underground injection of hazardous waste adjacent to known drinking water supplies. Finally, Congress was concerned that EPA's definition of hazardous waste had not included all materials which should be so designated.

In addition to these "loopholes", Congress was deeply concerned about the continued safety and wisdom of land disposal of hazardous waste. EPA regulations had allowed land disposal to continue. Thus, Congress itself, concerned about the growing body of evidence that land disposal is particularly hazardous, went about crafting a detailed and highly technical land disposal control regime. Moreover, the scandals and investigations about EPA's lack of enforcement of RCRA and Superfund prodded Congress to adopt wide-ranging amendments to bolster EPA's criminal and civil enforcement activities and to back them up by a broadened citizen suits provision. See generally, H. R. Rep. No. 198, 98th Cong., 1st Sess. (1983).

B. The 1984 Amendments

The 1984 amendments to RCRA were lengthy, complex, and highly technical. It is beyond the scope of this chapter to analyze them in detail. Thus, the following description is intended merely to highlight the major facets of this remarkable statutory revision.

The 1984 amendments required that EPA promulgate standards — no later than March 31, 1986 — for hazardous waste generated in quantities between 100 and 1000 kg/month. As a statutory "hammer" to force action by EPA, Congress also prescribed that, with certain exceptions, if EPA failed to promulgate the standards on time, hazardous waste generated in quantities greater than 100 kg/month would lose its exemption. Congress also required that during the interim, all "small quantity generator" wastes between 100 and 1000 kg would be subject to the uniform manifest system. EPA was also directed to study and report to Congress on the characteristics of the wastes generated by "small quantity generator". RCRA § 3001(d).

With respect to land disposal, Congress took two approaches; first, it banned the land disposal of certain wastes specifically listed in the statute unless EPA, within a prescribed time period, determined that the prohibition on such land disposal was not required in order to protect health and the environment. This prohibition on land disposal of certain specifically listed wastes would be effective 32 months after enactment. Congress went to extraordinary lengths to circumscribe EPA's power to find no health hazard. It required proof "to a reasonable degree of certainty, that there will be no migration" of the wastes while they remain hazardous. RCRA § 3004(d).

Second, Congress imposed specific land disposal restrictions with respect to certain bulk, noncontainerized liquid hazardous waste and the disposal of wastes in salt domes or through underground injection. RCRA § 3004(b), (c). Variances from the land disposal restrictions could only be granted — at least to extend the deadline — on the basis of the unavailability of alternative technology for disposal. RCRA § 3004(h).

Congress also tightened the requirements for so-called "interim status" surface impoundments for which RCRA permit applications were pending. RCRA § 3005(e), (j). Time limits were also imposed for the removal of hazardous treatment residues from surface impoundments that store or treat certain hazardous wastes banned from land disposals. RCRA § 3004(j).

Furthermore, new land disposal facilities were required by statute to have a double liner with a leachate collection system as well as an in-place ground water monitoring system. RCRA § 3004(o). Once permits are issued, they must be renewed every 10 years and permits for land disposal sites must be renewed every 5 years. RCRA § 3005(c). for land disposal sites must be renewed every five years. RCRA § 3005(c).

To anticipate public health hazards which may arise from such sites, Congress required that within 9 months all permit applications for landfills and surface impoundments be accompanied by an assessment of potential public hazards and a plan for "corrective action" to protect human health. RCRA § 3004(u), (v).

In order to stimulate the reduction in the quantity of hazardous waste generated through industrial processes, after September 1, 1985, all waste manifests were required to contain a certificate from the generator that the volume, quantity, and toxicity of the wastes had been reduced to the maximum degree "economically practicable." RCRA §§ 3002(b), 3005(h).

In order to force EPA to include additional substances as "hazardous waste", Congress required that EPA determine whether to list certain specifically identified wastes within time periods of 6, 12, or 15 months. RCRA § 3001(e). In order to clarify the standards for delisting such a hazardous waste, Congress specified that EPA consider factors in addition to those which caused the waste to be listed, if, for example, new data are available, and that EPA provide notice and an opportunity for public comment before making any decision on proposed delisting. RCRA § 3001(f).

To plug the "loophole" with respect to the burning of hazardous waste in boilers, Congress required EPA to promulgate detailed regulations, within prescribed time periods, governing persons who produce, burn, or distribute fuel derived from hazardous waste. The regulations were also required to include detailed recordkeeping requirements and, within 2 years of the enactment of the amendments, technical standards for the transportation of such fuel. RCRA § 3004(q). Warning labels were to be required immediately. RCRA § 3004(r). Within 12 months of enactment of these amendments, EPA was required to propose whether to list used automotive oil as a hazardous waste for which separate performance standards and manifest requirements would be announced. RCRA § 3014(b), (c).

Responding to concerns about EPA's enforcement efforts, Congress authorized EPA to assess civil penalties of up to $25,000 administratively for past and present violations of RCRA. RCRA § 3008(c). In addition, the maximum criminal penalties were increased and the list of conduct which qualifies for criminal sanctions was expanded. RCRA § 3008(d), (e). As a backup to EPA enforcement, Congress substantially broadened the scope of citizen suits permitted under the Act. With certain conditions, citizens were authorized to bring actions to curtail a past or present "imminent hazard". RCRA § 7002(a)(1)(B). Congress also clarified that the "imminent hazard" provisions of Section 7003 were intended to apply to past generators and past disposal sites, as well as to currently active sites. RCRA § 7003(a).

In a separate major new environmental action, Congress required EPA to establish regulations governing underground storage tanks. The underground storage tank requirements essentially created a new regulatory program. RCRA, Subtitle H, §§ 8001-8007.

This very brief discussion does not refer to all of the changes imposed by Congress through its 1984 amendments of RCRA. It does, however, serve to underscore the unusual nature of the legislative developments in hazardous waste which occurred in 1984.

IV. JUDICIAL INTERPRETATIONS OF RCRA

A. The Supreme Court

The Supreme Court has only tangentially addressed RCRA. In *Midlantic Nat'l Bank v. New Jersey Dep't of Envtl. Protection,* 106 S.Ct. 755 (1986) the Court recently held that because of the Congressional concern evidenced in RCRA and Superfund that hazardous wastes "may present an imminent and substantial endangerment to health or the environment," the Bankruptcy Court does not have the power to authorize an abandonment of contaminated property without formulating conditions that will adequately protect the public's health and safety. In that case, the trustee in bankruptcy had abandoned property in New York and New Jersey that was polluted by PCB-contaminated oil. The State of New York had decontaminated the property within its boundaries at a cost of $2.5 million.

In 1978 the Supreme Court held that RCRA does not preempt a New Jersey statute prohibiting importation of most "solid or liquid waste which originated or was collected outside the territorial limits of the state." *City of Philadelphia* v. *New Jersey,* 437 U.S. 617 (1978). The Court held that Congress did not expressly or implicitly intend to preempt the entire field of interstate waste management or transportation, but rather intended that the collection and disposal of solid wastes be primarily a function of the states. *Id.* at 620, n. 4. The Court went on, however, to hold the statute invalid as a violation of the Commerce Clause.

B. The Lower Courts

1. Private Enforcement Actions

The lower federal courts have required strict compliance with the statutory notice requirements of the citizen suits provision. The courts have held that notice to the EPA is a jurisdictional issue and have not hesitated to dismiss actions where the requisite notice was not given. See e.g., *Garcia* v. *Cecos Int'l, Inc.,* 761 F.2d 76 (1st Cir. 1984) (vacating the judgment of the district court where no actual notice was given to EPA); *Walls* v. *Waste Resource Corp.,* 761 F.2d 311 (6th Cir. 1985) (affirming dismissal); *Mola Development Corp.* v. *United States,* 22 Envtl. Rep. 1443 (C. D. Cal. 1985) (summary judgment granted); *Reeger* v. *Mill Service, Inc.,* 592 F. Supp. 1266 (W. D. Pa. 1984), and 593 F. Supp. 360 (W. D. Pa. 1984) (motions to dismiss granted).

The courts have also held that the citizen suits provision only allows for injunctive relief and cannot be used to recover damages. See *Walls* v. *Waste Resource Corp., supra.* Moreover, the provision does not allow a citizen to sue to force the EPA regional administrator to take enforcement action, since the regional administrator's enforcement authority is discretionary. *Proffitt* v. *Eichler,* 22 Envtl. Rep. 1106 (W. D. Pa. 1984).

2. Civil and Criminal Penalties

The Court granted partial summary judgment as to liability for civil penalties against a landfill operator in *United States* v. *Liviola,* 605 F. Supp. 96 (N. D. Ohio 1985). There, the operator failed to comply with an EPA request for information concerning the type, amount, and source of waste material transported to and disposed of at the site. The Court held that a party may be liable under the civil penalty provisions even if his actions were not willful, since the civil violation provisions impose strict liability, and even if the EPA has not yet issued a compliance order.

The Third Circuit has addressed the criminal violation provisions of RCRA that proscribe knowingly treating, storing, and disposing of hazardous wastes without a permit. *United States* v. *Johnson & Towers, Inc.,* 741 F.2d 662 (3rd Cir. 1984). The appellate court reversed a district court's decision holding that the criminal provisions only apply to owners and operators. The district court had held that the service manager and shop foreman of a truck repair company cannot be criminally liable because it found that Congress intended to limit criminal liability to persons who could have obtained a permit, and these employees could not have done so. Moreover, the court found that the defendants were not qualified to identify or analyze hazardous wastes. The Third Circuit rejected this interpretation of the criminal penalty provisions as unduly narrow and held that the provisions apply to all employees who knew or should have known that the waste was being disposed of without permits.

3. Issues in Bankruptcy

In addition to the landmark decision in *Midlantic Nat'l Bank, supra,* a district court in Texas has recently held that a debtor in bankruptcy is not entitled to an automatic stay of an impending order from EPA to file a permit application or closure plan for its hazardous waste facility. *In re Commonwealth Oil Refining Co., Inc.,* 23 Envtl. Rep. 1069 (W. D. Tex. 1985). The Court held that enforcement of environmental laws is the type of police or regulatory action exempted by Congress from the bankruptcy automatic stay provisions in 11 U.S.C. § 362(b)(4).

4. The Emergency Provisions of § 7003

The 1984 amendments expressly rejected two lines of cases emerging under § 7003, holding that: (1) private parties are not entitled to bring actions under the section and (2) the section does not authorize actions against past, non-negligent, off-site generators. See e.g., *United States* v. *Hooker Chemicals and Plastics Corp.,* 20 Envtl. Rep. 1857 (W. D. N. Y. 1984) (no private right of action); *United States* v. *A&F Materials Co., Inc.,* 582 F. Supp. 842 (S. D. Ill. 1984) (no action against past generators); *United States* v. *Northeastern Pharmaceutical and Chemical Co., Inc.,* 579 F. Supp. 823 (W. D. Mo. 1984) (same). The amendments expanded the scope of the citizen suits provision, which now provides a private right of action that parallels Section 7003. In addition, Section 7003 was made expressly applicable to past generators.

The Fourth Circuit has recently held that the owners and former operators of an inactive landfill may be sued under Section 7003 for their alleged failure to stop toxic contaminants from leaking into the ground-water. *United States* v. *Waste Industries, Inc.,* 734 F.2d 159 (4th Cir. 1984). The Court held that Section 7003 is not limited to emergency problems and that the term "disposal" as used therein applies to current conduct and encompasses such leaking.

Chapter 5

SUMMARY AND ANALYSIS OF THE RESOURCE CONSERVATION AND RECOVERY ACT OF 1976, AS AMENDED*

Richard deC. Hinds**

TABLE OF CONTENTS

*   All footnotes appear at the end of chapter.
** Partner, Cleary, Gottlieb, Steen & Hamilton, 1250 Connecticut Avenue, N.W., Washington, D.C. 20036. The assistance of Michael Wiegard and Janet Weller in preparing this article is gratefully acknowledged.

I. INTRODUCTION AND OVERVIEW

The Resource Conservation and Recovery Act of 1976, overview (RCRA or the "Act"), 42 U.S.C. §§ 6901-6987, as amended, authorizes the Environmental Protection Agency (EPA) to establish a comprehensive federal regulatory program to control the handling and disposal of hazardous wastes. The statute also calls for improved solid waste management by states and localities and mandates the phasing out of open dumps.

Passage of RCRA was prompted by Congressional concern that disposal of increasing amounts of solid waste had created serious problems for local communities, particularly those in urban areas; that disposal of solid and hazardous wastes without careful planning and management can be dangerous to human health and the environment; and that open dumping of wastes is particularly harmful to health, it contaminates drinking water supplies, and pollutes the air and the land (Section 1002(a), (b)).

The stated objectives of the Act are "to promote the protection of health and the environment and to conserve valuable materials and energy resources" by improving solid and hazardous waste management, prohibiting future open dumping, and promoting research, development, and demonstration programs (Section 1003).

The substantive provisions of the Act can be divided into three main categories: federal or state regulation of hazardous wastes; state control, pursuant to federal guidelines, of nonhazardous solid wastes; and federal programs in the areas of research and development, technical and financial assistance, and procurement. The approaches adopted within each of these categories are briefly summarized below.

The Federal Hazardous Waste Management Program

The Act directs the EPA to perform several tasks. First, it must identify those wastes which are hazardous either by name or by general characteristics. Second, it must issue standards for generators of hazardous wastes respecting record-keeping, labeling, containers, reporting, and use of the manifest system. A special manifest must accompany all hazardous wastes from point of generation to point of disposal. Third, EPA is required to issue standards applicable to transporters of hazardous wastes and to owners and operators of facilities engaged in treating, storing, or disposing of hazardous wastes. Finally, EPA is to issue regulations requiring all persons who treat, store, or dispose of hazardous wastes (including generators of such wastes) to obtain a permit to carry on such activities.

On May 19, 1980, EPA issued regulations establishing the framework for this hazardous waste program. The Act required all persons generating, handling, or disposing of hazardous wastes identified pursuant to these regulations to notify EPA of such activity by August 18, 1980. As mandated by RCRA, the hazardous waste regulations became effective on November 19, 1980 — 6 months after promulgation.

States with hazardous waste programs "substantially equivalent" to the Federal program will be granted interim authorization to carry out such programs for 2 years in lieu of the Federal program. A state may receive full authorization to administer and enforce its hazardous waste program only after developing a program which is "equivalent" to the federal program and consistent with other state hazardous waste programs.

State Solid Waste Management Programs

RCRA also directs EPA to develop and publish guidelines for solid waste management and provide minimum criteria for the development of state solid waste management plans. The states retain responsibility for administering and enforcing solid waste programs. However, federal financial assistance is available to states taking action to develop a plan which meets with federal approval. To be approved, such a plan must, among other things, prohibit the establishment of any new open dumps, and provide for the elimination of existing open dumps, either by closure or by conversion to sanitary landfills. The Act specifically prohibits disposal practices which constitute open dumping except where permitted under a state plan phase-out schedule.

Federal Programs to Reduce Solid Waste

The Act gives the federal government responsibility for several programs having the ultimate objective of reducing solid waste. The Department of Commerce is to promote the development of resource recovery technology. EPA is directed to engage in research, development, and demonstration projects in many areas of solid waste management, particularly solid waste utilization. Federal procurement policies are to favor the use of recycled material to the greatest extent consistent with maintaining satisfactory competition, performance standards, and a reasonable price. The Act also establishes an Interagency Coordinating Committee which is to submit a 5-year action plan to Congress and coordinate federal activities regarding resource conservation and recovery.

This chapter provides a detailed outline of the provisions of the Act which vitally affect anyone engaged in generating, storing, treating, transporting, or disposing of industrial wastes. As such it should serve as a guide to EPA's implementation of the Act over the next decade.

II. THE FEDERAL HAZARDOUS WASTE MANAGEMENT PROGRAM

The Act outlines a federal hazardous waste program with three basic components. First, EPA is to develop criteria for identifying and listing hazardous wastes (Section 3001(a)) and then issue regulations identifying characteristics of hazardous wastes and listing particular hazardous wastes by name (Section 3001(b)). Second, EPA is to promulgate standards applicable to generators of hazardous wastes, transporters of hazardous wastes, and owners and operators of hazardous waste treatment, storage, or disposal facilities (Sections 3002, 3003, and 3004, respectively). Third, EPA must establish a permit system applicable to owners or operators of facilities that treat, store, or dispose of hazardous wastes (Section 3005).

Under the original statutory time-table, this federal program (or substantially equivalent state programs) was to be fully implemented and effective by October 21, 1978. After EPA failed to meet this deadline, the Environmental Defense Fund and a number of other plaintiffs brought suit and obtained a court order requiring EPA to promulgate the required regulations according to a schedule drawn up by the Agency.[1] Pursuant to that schedule and order, EPA issued the regulations comprising the basic hazardous waste management program on May 19, 1980.[2]

A. The Definition of Hazardous Waste

The definition of "hazardous waste" is the key to the scope of the federal hazardous waste program. Waste materials that fall within the definition are subject to extensive regulation, those excluded are largely unregulated except perhaps at the state or local level. The Act defines "hazardous waste" as "a solid waste, or combination of solid wastes, which because of its quantity, concentration, or physical, chemical, or infectious characteristics may

1. Cause, or significantly contribute to an increase in mortality or an increase in serious irreversible, or incapacitating reversible, illness, or
2. Pose a substantial present or potential hazard to human health or the environment when improperly treated, stored, transported, or disposed of, or otherwise managed" (Section 1004(5))

Although hazardous waste is a subcategory of "solid waste", RCRA includes within the definition of solid waste a number of wastes which might not be considered too "solid". In addition to garbage, refuse, or sludge, the definition covers "other discarded materials, including solid, liquid, semisolid, or contained gaseous material resulting from industrial, commercial, mining and agricultural operations . . . ." Specifically excluded from the definition of solid waste are domestic sewage wastes, industrial effluent discharges subject to permits under Section 402 of the Clean Water Act, and specified nuclear materials subject to regulation under the Atomic Energy Act (Section 1004(27)).

Although the Act defines solid waste as "discarded material" (Section 1006(27)), EPA has taken the position that reused or reclaimed materials are subject to regulation as hazardous waste if they are "sometimes discarded",[3] and has placed restrictions on the storage and transportation of such materials. Whether the Act authorizes EPA to regulate materials which are in fact not discarded by a particular producer remains an open question. The Hazardous and Solid Waste Act Amendments ("the Amendments") require EPA to expand the universe of hazardous wastes significantly.

B. Identification and Listing of Hazardous Wastes

The Act requires EPA, after notice and opportunity for public hearing, to develop criteria which will enable it to identify the characteristics of hazardous wastes and to list specific wastes as hazardous. In establishing such criteria, EPA is instructed to take into account "toxicity, persistence, and degradability in nature, potential for accumulation in tissue," as well as flammability, corrosiveness, and other hazardous characteristics. The hazardous waste criteria are to be revised from time to time as may be appropriate.

Using these hazardous waste criteria, EPA is directed to identify the general characteristics of hazardous waste and list particular hazardous wastes. The criteria establish the standard of judgment, while the regulations reflect the applications of the criteria to specific wastes.[4] Congress anticipated the regulations would identify two basic types of substances: those which are hazardous in themselves and those which are hazardous when present in sufficient quantity or concentration or when mixed with other substances.[5]

EPA published initial lists of specific hazardous wastes in its May 19, 1980 regulations and designated ignitability, reactivity, corrosivity, and extraction procedure toxicity as the four characteristics of hazardous waste.[6] The criteria for listing wastes as hazardous include meeting any of the four designated characteristics, exhibiting certain acutely hazardous properties, or containing certain toxic constituents.

The 1980 amendments to the statute[7] exempted the following wastes from the hazardous waste program pending the completion of studies of their environmental effects:

1. Drilling fluids, produced waters, and other wastes associated with the exploration, development, or production of crude oil or natural gas or geothermal energy;
2. Fly ash waste, bottom ash waste, slag waste; and flue gas emission control waste generated primarily from the combustion of coal or other fossil fuels;
3. Solid waste from the extraction, beneficiation, and processing of ores and minerals; and
4. Cement kiln dust waste

Any regulation of these wastes as hazardous waste must not occur until at least 6 months after submission of EPA studies on the effects of these wastes on health and the environment. Any regulations covering drilling fluids and other oil and gas exploration or production wastes would, in addition, not become effective until affirmatively authorized by an Act of Congress. EPA has 2 years to complete its reports on the effects of all these wastes except cement kiln dust, for which the statute allows a 3-year study period (Section 3001(b)).

The Act contains procedures by which states and other interested persons can seek changes in the regulations identifying and listing hazardous wastes. A governor of a state may petition EPA to identify or list additional wastes as hazardous. The Agency has 90 days to act upon such a petition (Section 3001(c)). The Act also authorizes any person to petition for the promulgation, amendment, or repeal of any regulation issued under the Act, including regulations identifying or listing certain substances as hazardous wastes. EPA must act on such a petition within a reasonable time and publish the reasons for its decision (Section 7004(a)).

C. Hazardous Waste Notification

The Act required all persons generating, transporting, or treating, storing, or disposing of wastes identified or listed as hazardous in EPA regulations to file a notification with the Agency within 90 days after the May 19, 1980 regulations were promulgated. The notification had to include a general description of the regulated activity, its location, and a list of all hazardous wastes being handled. Persons not handling identified or listed hazardous wastes at the time the regulations were published were not required to notify EPA. Any person handling hazardous wastes who failed to properly notify EPA of his activities cannot legally transport, treat, store, or dispose of such wastes after the hazardous waste management program became effective on November 19, 1980.

Under the 1980 amendments, EPA has the discretion to require the submission of new notifications when it revises or adds to Section 3001 characteristics or waste listings, but such notifications are no longer automatically required.

D. Standards Applicable to Generators of Hazardous Waste

The Act also requires EPA to promulgate standards applicable to generators of hazardous waste.[8] EPA's standards, which took effect on November 19, 1980, set forth the hazardous waste management practices which generators must follow.[9]

The Act specifies that the standards for generators of hazardous wastes are to establish such requirements "as may be necessary to protect human health and the environment" respecting:

1. Record-keeping
2. Labeling
3. Use of appropriate containers
4. Furnishing of information on the general chemical composition of the wastes to persons further down the waste stream

5.   Use of a manifest system and any other reasonable means necessary to insure that all hazardous wastes are sent to and arrive at facilities which have the necessary permits[10]

6.   Reporting to EPA on the quantities of wastes generated and their disposition

Congress envisioned that rather than restricting the generation of hazardous waste, which in many instances would interfere with the productive process itself, the responsibility of the generator under EPA's regulations would be generally limited to one of providing information.[11]

The RCRA regulations (40 C.F.R. Part 262) issued by EPA of necessity go beyond record-keeping and reporting. First and foremost, they impose on generators of solid waste the obligation to determine whether it is a hazardous waste. Generators must also properly package hazardous wastes prior to transport off-site and must comply with the manifest system for tracking hazardous waste shipments. Any generator who stores hazardous wastes on-site for more than 90 days, or who treats or disposes of hazardous wastes on-site must also comply with the regulations applicable to owners and operators of hazardous waste treatment, storage, and disposal facilities.[12] Finally, generators of hazardous wastes must maintain specified records and file annual and other reports.

In the belief that at least initially the principal focus of the regulatory program should be on the larger generators of hazardous wastes, EPA's regulations contain a two-tiered exemption for generators who produce only small quantities of such wastes.[13] These provisions set a general exemption level of 1000 kg/month per site, with lower levels for wastes which are "acutely hazardous".[14] The Amendments require EPA to regulate generators of more than 100 kg of hazardous waste.

E. Standards Applicable to Transporters of Hazardous Wastes

The standards for transporters parallel those for hazardous waste generators. The Act directs EPA to promulgate regulations establishing such standards "as may be necessary to protect human health and the environment", including but not limited to standards on:

1.   Record-keeping
2.   Enforcement of labeling requirements
3.   Compliance with the manifest system (Section 3003(a))

The standards for hazardous waste transporters apply whenever the hazardous waste regulations require use of a hazardous waste manifest. The RCRA regulations require transporters of hazardous waste to obtain identification numbers, comply with the manifest system, and report and clean-up any discharges that occur during transportation.[15]

The Act requires that any standards issued by EPA for the transportation of hazardous wastes which are also subject to the Hazardous Materials Transportation Act, 49 U.S.C. §§ 1801 *et seq.,* must be consistent with that Act and any implementing regulations issued by the Department of Transportation (DOT) (Section 3003(b)). DOT has decided to regulate all hazardous wastes as hazardous materials, and EPA in turn has simply adopted by reference DOT regulations regarding labeling, marketing, placarding, packaging, container requirements, and reporting of spills.[16]

F. Standards for Hazardous Waste Treatment, Storage, and Disposal Facilities

The Act grants EPA comprehensive authority to regulate the management of hazardous wastes by persons who own or operate "hazardous waste treatment, storage, or disposal facilities" (TSDFs). The terms "treatment", "storage", and "disposal" are broadly defined in the Act (Section 1004(34), (33), and (3), respectively). *Treatment* includes neutralization, evaporation, or *any other change* in the character or composition of a hazardous waste. EPA has taken the position in its regulatory definition of treatment that the term covers recycling and recovery operations.[17] *Storage* includes the temporary containment of hazardous waste. *Disposal* includes the discharge, deposit, injection, dumping, spilling, or leaking of any solid or hazardous waste into or on any land or water.[18]

Neither the Act, its legislative history, nor the implementing regulations make a distinction between persons who operate hazardous waste facilities in connection with manufacturing activities and persons who operate such facilities independently of such activities. Thus, a generator of hazardous waste who also treats and stores such waste would be required to comply with the relevant standards and to obtain the necessary permits.

The Act requires EPA to promulgate performance standards for owners and operators of hazardous waste facilities "necessary to protect human health and the environment" including but not limited to the following:

1. Record-keeping
2. Reporting and compliance with the manifest system
3. Operating methods, techniques, and practices
4. Location, design, and construction of hazardous waste facilities
5. Contingency planning
6. Operational practices, continuity of operation, and financial responsibility, and
7. Compliance with the Section 3005 permit system (Section 3004)

Section 3004 of the Act further specifies that no private entity should be precluded by EPA's financial responsibility requirements from owning or operating hazardous waste facilities, where such entity can provide assurances of financial responsibility and continuity of ownership consistent with the degree of risk associated with the operation of the facility.

Section 3004 of RCRA was amended in 1980 to require that, in issuing TSDF standards, EPA shall "where appropriate" distinguish between new facilities and facilities in existence on the date such standards are promulgated. The legislative history indicates such a distinction would be "appropriate" where practical and technical limitations make it unduly burdensome to require that existing facilities be modified or retrofitted. Waste-water treatment impoundments were mentioned as a specific example.[19]

EPA is developing the standards for hazardous waste management facilities in three phases. The issuance on May 19, 1980 of standards for TSDFs with "interim status" constituted Phase I of this program. The Phase II regulations are to establish final standards to be used in processing TSDF applications for permits under the Act. Some Phase II regulations have been issued and can be found in 40 C.F.R. Part 264 but important segments (e.g., land disposal standards) are still being developed. Interim standards for new hazardous waste land disposal facilities can be found in 40 C.F.R. Part 267. In Phase III, EPA plans to issue more definitive counterparts of the Phase II standards. The entire process of finalizing the regulations and granting the first round of permits to TSDFs is expected to take a minimum of 5 to 6 years from the completion of Phase II.

G. The Permit Program for Hazardous Waste Facilities

In addition to requiring EPA to issue performance standards for hazardous waste facilities, the Act directs the Agency to issue regulations creating a permit system for such facilities (Section 3005(a)). Congress recognized that the issuance of permits to TSDFs could involve a considerable period of time. It therefore provided in Section 3005(e) of RCRA that a hazardous waste facility could obtain "interim status" to operate pending final administrative disposition of the permit application for the facility. A facility which has interim status is deemed to have a permit until EPA takes final action on its permit application.

In order to be eligible to obtain interim status, a hazardous waste facility must have been "in existence" on November 19, 1980.[20] In addition, the facility's owner or operator must have complied with the notification requirements of Section 3010(a) (to the extent applicable) and submitted a RCRA permit application. Facilities not required to file notifications under Section 3010 will be deemed to have "complied" with its requirements and thus meet the first prerequisite for interim status.[21] EPA has also taken the position that existing facilities not subject to the hazardous waste regulations as of November 19, 1980 can obtain interim status by filing a permit application within 6 months of an amendment to the regulations which first subjects the facility to the RCRA hazardous waste regulations or within 30 days after it loses a regulatory exemption.[22]

Both the standards for hazardous waste facilities with interim status[23] and the RCRA permit program[24] took effect on November 19, 1980. Therefore, the treatment, storage, or disposal of hazardous waste without either interim status or a permit is now illegal (Section 3005(a)), and can result in the imposition of civil and criminal penalties (Section 3008(a)(3), (d)(2), (e)(1)(A)). A facility with interim status must operate in compliance with the interim status standards in Part 265 of the RCRA regulations until action is taken on its permit application. A facility that is issued a permit must operate in compliance with the terms of its permit and the applicable Phase II facility standards.[25] If the permit application for a facility is denied, the nonpermitted hazardous waste activity must cease.

The Act specifies that EPA is to require the submission of at least the following information in permit applications:

1. Estimates as to the quantity, composition, and concentration of hazardous waste, or combinations of such waste with other solid waste, proposed to be treated, stored, transported, or disposed of;
2. The time, frequency, or rate of treatment, storage, transportation, or disposal of such wastes; and
3. The site at which such wastes will be treated, stored, transported to, or disposed of.

The application must also establish that the facility complies with the applicable hazardous waste facility standards. If the applicant indicates in its permit application that modification of a facility will be necessary to ensure compliance, the permit must establish a time to complete such modification (Section 3005(c)).

EPA has adopted a two-step permit application procedure. Facilities desiring interim status must file a relatively brief form (Part A of the RCRA permit application) setting forth basic information on the nature and location of the facility. Thereafter, facilities will be selected for permitting on a prioritized basis and requested to file a more extensive, narrative application (Part B) demonstrating compliance with the final TSDF standards in Part 264 of the RCRA regulations.

Public notice of the intent to issue a facility a TSDF permit must be given and an informal hearing held if requested by a person opposing issuance of such a permit or if deemed appropriate by the permitting authority. The hearing must be held prior to issuance of such a permit by either EPA or a state (Section 7004(b)(2)).

H. State Hazardous Waste Programs

The Act requires EPA to promulgate guidelines to assist states in developing their own hazardous waste programs (Section 3006(a)).[26] It appears that most states will seek authority to administer their own hazardous waste programs and several have already received interim authorization. The Act provides for two different types of EPA approval of state hazardous waste programs — interim authorization and final authorization.

A state that has a hazardous waste program in existence 90 days after issuance of the May 19, 1980 RCRA regulations may seek interim authorization from the Administrator to carry out that program in lieu of the federal program. Upon a determination that the state program is "substantially equivalent" to the federal program, EPA may grant such authorization for a 2-year period commencing on November 19, 1980, the effective date of the federal hazardous waste program (Section 3006(c)).

The Agency, after providing notice and opportunity for public hearing, is required to give final authorization to a state program unless it finds:

1.   The program is not equivalent to the federal program;
2.   The program is not consistent with federal or state programs applicable in other states; or
3.   The program does not provide adequately for enforcement (Section 3006(b)).

If EPA determines, following public hearing, that a state is not properly administering and enforcing its program, the Agency may withdraw authorization of the state program if the state fails to take appropriate corrective action within 90 days (Section 3006(e)).

I. Hazardous Waste Site Inventory

The 1980 Amendments added a provision to RCRA which requires each state to undertake, "as expeditiously as practicable," a continuing program to compile, publish, and submit to EPA an inventory describing the location, ownership, contents, and condition of each site within its borders at which hazardous waste has been stored or disposed of (Section 3012(a)). Any state may order any person to compile this inventory. EPA may make grants to the states to carry out such a program (Section 3012(c)).

If the Agency determines that a state inventory program is not adequately providing the required information, it is empowered to take over the program (Section 3012(b)). The Act expressly provides that enforcement or remedial actions need not be postponed pending completion of the required hazardous waste site inventory (Section 3012(d)).

J. Monitoring, Analysis, and Testing

Upon determining that an active or inactive hazardous waste treatment, storage, or disposal site may present a substantial hazard to human health or the environment, EPA may order the site owner or operator to conduct monitoring, testing, and analysis to ascertain the nature and extent of such hazard and to file reports thereon (Section 3013(a)). In the case of an inactive facility, if EPA finds that the current owner could not reasonably be expected to have actual knowledge of the presence of hazardous waste at the site and of its potential for release, the Agency may order the most recent previous owner or operator who could reasonably be expected to have such knowledge to carry out the monitoring, testing, and reporting (Section 3013(b)). A person who receives such an order has 30 days to submit a proposal for carrying out its requirements. If EPA determines that the owner or operator is unable to satisfy its requirements, EPA may conduct the monitoring and testing itself and seek reimbursement from the owner or operator for the costs incurred (Section 3013(d)). EPA also may commence a civil action seeking a civil penalty of up to $5000/day for failure to comply with a monitoring, analysis, or testing order (Section 3013(e)).

K. Federal Enforcement
1. Compliance Orders

Whenever EPA determines that a person is in violation of any requirement of the federal hazardous waste program, EPA may issue a compliance order or commence an action in federal district court for injunctive and other "appropriate relief" (Section 3008(a)). Under Section 3008(a), as amended in 1980, EPA may issue a compliance order as soon as a determination is made that a person is violating a requirement of the hazardous waste program. The compliance order may either require immediate compliance or specify a time period for coming into compliance.

A compliance order is required to state the nature of the alleged violation "with reasonable specificity" and to specify a time for compliance. The order may also assess a penalty "which the Administrator determines is reasonable, taking into account the seriousness of the violation and any good faith efforts to comply with the applicable requirements" (Section 3008(c)).

A compliance order becomes final 30 days after service unless the recipient, in the interim, requests a public hearing. Upon receiving such a request, EPA is directed to conduct such a hearing promptly (Section 3008(b)). A person who fails to take corrective action is liable for a civil penalty of up to $25,000 for each day of continued noncompliance. In addition, EPA may suspend or revoke any permit issued to the violator, whether issued by the Agency or the state (Section 3008(a)(3)).

If the violation occurs in a state with an authorized hazardous waste program in effect, EPA is required to give the state notice before taking any enforcement action. If the state takes apropriate action against the violator, the Agency presumably would not initiate any enforcement action. If the state fails to take such action, EPA may proceed to issue a compliance order or commence a civil action (Section 3008(a)(2)).

2. Permit Suspension or Revocation

If EPA (or a state with an authorized hazardous waste program) determines that a permitted facility is not in compliance with the terms of its permit or with applicable standards for hazardous waste facilities, the Agency (or state) is required to revoke the permit (Section 3005(d)). Such a revocation or suspension order becomes final 30 days from issuance unless the recipient in the interim requests a public hearing (Sections 3008(b), (c)). Any permit suspension or revocation imposed by EPA following such a hearing would be subject to judicial review under the Administrative Procedure Act (5 U.S.C. §§ 701 *et seq.*).

3. Civil Penalties

Any person who violates one of the hazardous waste management provisions of the statute is subject to a civil penalty of not more than $25,000 for each violation (Section 3008(g)). Each day that a violation continues is considered a separate violation for purposes of assessing the penalty.

4. Criminal Sanctions

The Act also provides criminal penalties for persons who *knowingly:* (1) transport hazardous waste to a facility which does not have a permit; (2) treat, store, or dispose of hazardous waste without a permit or in violation of a material permit condition or requirement of a permit; (3) make a material false statement in any application, label, manifest, record, report, permit, or other document required by the hazardous waste program; or (4) destroy, alter, or conceal records required to be maintained under the hazardous waste program (Section 3008(d)). The first offense subjects the violator, upon conviction, to a fine of up to $25,000 for each day of the violation and/or imprisonment for up to 1 year ($50,000/day and 2 years imprisonment for transporting hazardous waste without a permit or for treating, storing, or disposing of wastes without a permit or in violation of a permit). A second violation is punishable by a fine of up to $50,000/day and/or imprisonment for up to 2 years.

The 1980 amendments to RCRA added a new provision to the Act which permits more severe criminal penalties to be imposed for certain life-threatening conduct. A person who *knowingly* violates specified statutory or regulatory requirements may be convicted of the new offense of "endangerment" if he also "knows at that time that he thereby places another person in imminent danger of death or serious bodily injury . . . ." (Section 3008(e)).

To warrant conviction of this new crime, the violator's conduct must manifest either (1) "an unjustified and inexcusable disregard for human life" or (2) "an extreme indifference for human life" (Section 3008(e)(2)). A person convicted under this provision can receive a fine of up to $250,000 or up to 5 years imprisonment, or both. Imprisonment for up to 2 years is authorized for conduct found to manifest "an unjustified and inexcusable *disregard* for human life." A term of up to 5 years can be imposed for conduct which manifests "*extreme indifference* for human life" (emphasis added). Upon conviction a defendant that is an organization is subject to a fine of not more than one million dollars.

The Act specifies that a person has committed a "knowing" act of endangerment if he is aware or believes that his conduct is substantially certain to cause danger of death or serious bodily injury (Section 3008(f)). The statute goes on to provide that a defendant who is a natural person cannot have attributed to him the knowledge of another person which the defendant himself did not actually possess. In proving the defendant's possession of actual knowledge, however, circumstantial evidence may be used, including evidence that the defendant took affirmative steps to shield himself from relevant information. The criminal responsibility of a corporation for knowledge possessed by its officers and managing agents is governed by traditional legal principles.

The Act states that all defenses available against criminal prosecution for other federal offenses may be raised in endangerment cases. In addition, a defendant can defend by showing that the danger and conduct charged were reasonably foreseeable occupational hazards to which the endangered person consented. Furthermore, the new provision expressly authorizes judicial development of defenses of justification and excuse.

The purpose of this provision is to establish enhanced felony penalties for certain life-threatening conduct.[27] The legislative history makes clear that actual harm is not an element of the crime.[28] Because no concrete harm need result for a person to be prosecuted and convicted, Congress was careful to circumscribe the reach of the new offense of endangerment by adding a precisely drawn *scienter* or intent requirement. Congress expressly disclaimed any intent to have persons "prosecuted or convicted unjustly for making difficult business judgments" without the necessary *scienter.*[29]

5. Imminent Hazards

The Administrator is authorized to bring suit in federal district court to immediately restrain any person from handling a solid or hazardous waste which "may represent an imminent and substantial endangerment to health or the environment" (Section 7003). This provision is in addition to, and independent of, the other provisions of the Act. EPA, through the Department of Justice, filed more than 50 such suits during 1979 and 1980. One of the first judicial rulings under Section 7003 indicated that the provision is jurisdictional in nature and does not eliminate the need to show irreparable harm in order to obtain preliminary injunctive relief.[30] Another early decision held that an allegation of ongoing acts of disposal is not essential in a Section 7003 action.[31] The court also ruled that Section 7003 applies to acts which occurred prior to the enactment of RCRA.[32]

In addition to going to court, Section 7003, as amended in 1980, also empowers the Administrator, after giving notice to the affected state, to issue such orders "as may be necessary to protect public health and the environment." EPA may sue in federal district court to recover fines of up to $5000 for each day for which a willful violation or failure to comply with such an order continues. The Agency may also use its inspection authority under Section 3007 in conjunction with enforcement actions taken under Section 7003 (see discussion below).

III. STATE OR REGIONAL SOLID WASTE PLANS

A. Introduction

While the hazardous waste program has been developed and, in many states, is administered and enforced by EPA, the nonhazardous solid waste program established by the Act is administered and enforced solely by the states. EPA's role is limited to providing the states with technical and financial assistance to encourage them to engage in comprehensive planning pursuant to federal guidelines.

Congress rejected federal preemption of state solid waste programs as "undesirable, inefficient, and damaging to local initiative."[33] However, the Act contains a strong inducement to the development of state solid waste plans: a prohibition against waste disposal practices which constitute open dumping. The Act provides that where alternatives to open dumping are not available, such practices can continue only if a state solid waste plan contains a schedule requiring the entity engaging in such dumping to comply with the prohibition on open dumping within a reasonable time but at least by September 13, 1984 (Section 4005(a)).[34]

B. Federal Guidelines
1. Guidelines for State Plans

EPA is required to publish guidelines identifying those areas of the country which have common solid waste problems appropriate for regional planning and management (Section 4002(a)).[35] The Agency must, in addition, promulgate guidelines to assist in the development and implementation of state solid waste management plans ("state plans") (Section 4002(b)). The Act requires these guidelines to reflect consideration of a comprehensive list of factors, including regional, geologic, hydrologic, and climatic characteristics; relevant political, economic, and organizational problems; the industrial profile; and the constituents and generation rates of waste (Section 4002(c)). In addition, these guidelines are to "consider . . . methods for closing or upgrading open dumps for purposes of eliminating potential health hazards" (Section 4002(c)(3)).[36]

2. Solid Waste Management Guidelines

EPA is also required to publish suggested guidelines for solid waste management. Section 1008 of the Act states that, among other things, these guidelines are to:

1.   Describe the available waste management practices (Section 1008(a)(1))

2.   Describe the minimum levels of performance appropriate to protect public health and welfare, the environment, including ground water, and aesthetic considerations (Section 1008(a)(2))
3.   Provide criteria to identify practices which constitute open dumping (Section 1008(a)(3))[37]

C. State Plans

The Act identifies the minimum elements a state plan must include to receive federal approval (Sections 4003, 4007(a)). They are

1.   An identification of the responsible government units and their regulatory powers
2.   A prohibition of new open dumps
3.   A requirement that all solid waste shall be utilized for resource recovery, disposed of in sanitary landfills, or otherwise disposed of in an environmentally sound manner
4.   A plan to close or upgrade all existing open dumps
5.   A provision for revising the state plan where EPA determines that such revisions are necessary

Within 6 months of submission of a state plan, EPA must approve the state plan if it contains these minimum requirements. The Agency may withdraw approval of a state plan upon a determination, after notice and opportunity for a public hearing, that revision or correction is necessary to bring the plan into compliance with the minimum requirements established under Section 4003.

EPA is authorized to provide financial assistance to the states for the development and implementation of waste programs. A state's eligibility for continuing financial asssistance is dependent on timely development of a state plan and subsequent approval of the plan by EPA (Section 4007(b)(1)). Such assistance can be provided for items such as expert consultants and feasibility studies, but may not be used for land acquisition, or any price subsidies for recovered resources (Section 4008(a)). The Act specifies the manner in which financial assistance is to be allotted among the states, and within a state (Section 4008(b), (c)). Special provisions are made for local governments and small or rural communities with serious solid waste disposal problems (Sections 4008(3), 4009(a)).

D. Conversion of Open Dumps to Sanitary Landfills

One of the stated objectives of the Act is to prohibit the future establishment of new open dumps and to convert existing open dumps to sanitary landfills which do not pose a danger to the environment or to health (Section 1003(3)).[38] Thus, state plans must, at a minimum, provide for the closing of open dumps or their upgrading to sanitary landfills and prohibit the establishment of new open dumps (Section 4003(2), (3)).[39] The means by which the closure or upgrading of existing dumps is to be accomplished are generally left to the states, although EPA has required that the state plan include some type of permitting mechanism in this regard. The legislative history and the statutory provisions for phasing out activities which constitute open dumping over 5 years (Section 4005) suggest that Congress envisioned the completion of this task over a 5-year period following EPA's promulgation of a national inventory of open dumps.[40]

1. Federal Criteria and Inventory of Open Dumps

An open dump is any facility or site which is not a sanitary landfill or an approved facility for the disposal of hazardous waste (Section 1004(14)). EPA must promulgate regulations that establish criteria for determining which facilities are sanitary landfills and which are open dumps (Section 4004(a)).[41] At a minimum, the criteria may classify a facility as a sanitary landfill "only if there is no reasonable probability of adverse effects on health or the environment from disposal of solid waste at such facility." *Id.* EPA must also publish an inventory of all disposal facilities in the U.S. which are open dumps (Section 4005(b)). State solid waste management plans must require that all disposal facilities listed in this inventory "comply with such measures as may be promulgated by the Administrator to eliminate health hazards and minimize potential health hazards" (Section 4005(a)).

2. Prohibition Against Open Dumping

In addition to requiring that each State plan prohibit the establishment of new open dumps and provide for the closing or upgrading of existing open dumps, the Act contains a prohibition against the practice of open dumping. Section 4005(a) provides that the open dumping of solid or hazardous waste is prohibited upon the promulgation of criteria under Section 1008(a)(3) defining what constitutes open dumping. These criteria were promulgated on September 13, 1979, and are codified at 40 C.F.R. Part 257. Congress recognized that there may not be any available alternatives for certain entities which dispose of their wastes through open dumping. Section 4005(a) therefore provides that a State plan may include a schedule for phasing out such disposal practices over a 5-year period commencing with promulgation of the federal open dumping criteria if the entity has considered alternatives but is unable to utilize them. Disposal practices which do not meet these criteria and thus constitute open dumping are prohibited unless and until the waste disposer receives an extended compliance schedule under a state solid waste management plan.

The Act does not provide for federal enforcement of the prohibition against open dumping. Enforcement is instead left to litigation brought by states or private parties under the Act's citizen suit provision (Section 7002).

There is no statutory provision requiring states to develop state plans and no provision for implementation of a federal solid waste program in states without such plans. Where a state has not yet prepared or received federal approval for the relevant portions of its solid waste mangement plan, obtaining the temporary exemption from the open dumping prohibition which the Act authorizes to be included in state plans apparently will not be possible. An entity whose disposal practices can be characterized as open dumping according to the criteria promulgated under Section 1008(a)(3) will thus be vulnerable to citizen suits until a state plan which includes a phase-out schedule for its disposal activities is approved by EPA. In an attempt to rectify this situation, EPA has indicated that it will expedite approval of portions of state plans which provide schedules for compliance with the statutory prohibition against open dumping.[42]

Although EPA views the open dump inventory called for by Section 4005(b) as merely an informational tool and adjunct to the state planning process,[43] listing a facility as an open dump in many cases will be tantamount to a determination that any disposal activities at that facility constitute open dumping. If so, persons disposing of wastes at listed facilities will be vulnerable to citizen's suits seeking to enforce the statutory prohibition against open dumping unless the facility is covered by a phase-out schedule in a state plan.

E. Imminent Hazards

The "imminent hazard" provision of the Act (discussed in Section II.K.5, *supra*) applies to both solid and hazardous wastes. EPA is authorized to bring suit in federal district court to immediately restrain any person from handling a solid waste if it "may present an imminent and substantial endangerment to health or the environment" (Section 7003). Thus, it is unnecessary for the Agency to have designated a particular solid waste as hazardous before seeking injunctive relief from the courts under Section 7003.

IV. ADMINISTRATIVE AND MISCELLANEOUS PROVISIONS

A. Coordination with Other Laws

The Act does not apply to "any activity or substance" subject to certain specified laws, including the Clean Water Act, except insofar as application of RCRA would not be inconsistent with the requirements of the other laws (Section 1006(a)). EPA is also instructed to avoid duplication by integrating RCRA with other environmental legislation granting it regulatory authority to the extent such integration can be done consistently with the goals and policies of such laws (Section 1006(b)).

The Act was amended in 1980 to specify that the Secretary of the Interior shall have exclusive responsibility for carrying out the federal hazardous waste management program for coal mining wastes and overburden for which a permit is issued under the Surface Mining Control and Reclamation Act of 1977 (Section 1006(c)(2)). The Act provides that the Secretary of the Interior may promulgate regulations in this regard "with the concurrence of" EPA. *Id.* The statute also now provides that a permit for the treatment, storage, or disposal of coal mining wastes and overburden issued under the Surface Mining Act is to be deemed a permit for RCRA purposes as well (Section 3005(f)).

While Section 1006(a) excludes from the scope of the Act the industrial discharge of wastes regulated under the Clean Water Act, EPA has taken the position that RCRA's requirements do apply to industrial wastewater being stored or treated prior to discharge.[44] Industry representatives have strongly objected to this assertion of regulatory jurisdiction, viewing the imposition under RCRA of new standards and permit requirements on such wastewater treatment facilities as inconsistent with the requirements and policies of the Clean Water Act and therefore precluded by Section 1006. The 1980 amendments to the Act did not explicitly address the issue of subjecting wastewater treatment ponds to regulation under RCRA. Section 3004 was amended, however, to provide that the Administrator shall, "where appropriate", distinguish between new and existing TSDFs in establishing facility standards.

B. Inspection

For the purpose of developing regulations or enforcing *any* of the provisions of RCRA, any person who handles or has handled hazardous wastes is required to permit EPA, upon request, to have access to and copy *all* records relating to such wastes (Section 3007(a)(1)). This inspection authority enables "duly designated" EPA officers, employees, and representatives to enter and inspect places where hazardous wastes are or have been generated, stored, treated, or disposed of, and to obtain samples (Section 3007(a)(1), (2)). The legislative history of the 1980 amendments indicates that Congress intended this authorization to include the use of private contractors.[45] If samples are obtained, the owner or operator (or his agent) may request a portion equal to the portion taken. If EPA analyzes the sample taken, a copy of the test results must be furnished promptly to the owner or operator. Information so obtained is publicly available, subject to a claim of business confidentiality by any person (Section 3007(b)).

C. Citizen Suits

Any person may bring a civil action in federal district court against any person alleged to be in violation of any requirement arising under the Act or against the EPA Administrator for his failure to discharge a nondiscretionary duty (Section 7002(a)).[46] Prior to bringing an action against a private party, the plaintiff must provide 60 days prior notice to EPA, the state in which the alleged violation occurred, and the alleged violator (Section 7002(b)(1)). Sixty days prior notice to EPA is also required before suing the Administrator (Section 7002(c)). An action against a private party is barred if EPA or the state is diligently prosecuting a similar action against the same party; however, any person may intervene as of right in such a suit (Section 7002(d)). A savings provision explicitly preserves any other statutory or common law rights a plaintiff may have to enforce solid or hazardous waste management standards or requirements, or to seek other relief against EPA or a state agency (Section 7002(f)).

D. Citizen's Petitions

Any person may petition EPA to promulgate, amend, or repeal any regulation under this Act. Within a "reasonable time" following receipt of such a petition, the Agency is required to act on the petition and publish notice of such action in the *Federal Register,* together with the reasons thereof (Section 7004(a)).

E. Judicial Review

Judicial review of regulations promulgated by EPA under RCRA or the Agency's denial of a petition for regulatory action must be sought within 90 days of the date of such promulgation or denial in the U.S. Court of Appeals for the District of Columbia Circuit (Section 7006(a)(1)).

The issuance, denial, modification, or revocation of a RCRA permit and the granting, denial, or withdrawal of authorization or interim authorization of a state hazardous waste management program by EPA is reviewable in the Court of Appeals for the circuit in which the petitioner resides or transacts business (Section 7006(b)). Such review must be sought within 90 days of the EPA action unless the application for review is based solely on grounds which arose after the 90th day.

F. Employee Protection

Section 7001 protects employees against discrimination for bringing or participating in a proceeding under the Act and authorizes the Department of Labor to enforce this provision (Section 7001(b)). Any employee threatened with loss of employment because of the administration or enforcement of the Act may request that Department to investigate the matter. At the request of any party, the Department must conduct public hearings on the record, and make findings of fact and recommendations (Section 7001). EPA is required to provide information to the Department identifying the hazards to which workers handling hazardous wastes at a particular facility may be exposed (Section 7001(f)).

V. FEDERAL PROGRAMS TO REDUCE SOLID WASTE

The Act gives the Department of Commerce, EPA, and a new Interagency Coordinating Committee responsibility for conducting a variety of programs to encourage solid waste utilization and reduction.

A. Duties of the Department of Commerce

The Department of Commerce is charged with the responsibility for encouraging greater commercialization of resource recovery technology. By September 1, 1981, the Department is required to publish guidelines for use by industry and government in preparing specifications for recovered materials (Section 5002).[47] Within this same period, the Department is also instructed to take the following actions:

1.   Identify markets for recovered material;
2.   Identify barriers to the use of recovered material; and
3.   Encourage new uses of recovered materials (Section 5003).

Additionally, the Department is authorized to evaluate the feasibility of resource recovery facilities and to develop a data base on the subject (Section 5004). In establishing any policies or controls which affect the development of markets for recovered materials, the Department is instructed to consider whether to establish the same or similar policies or controls for virgin materials (Section 5005).

B. Duties of EPA

The Act directs EPA to conduct and promote research and development by others in a number of waste management related areas (Section 8001(a)). In addition, the Agency is instructed to conduct a number of specific studies, including analyses of the problems connected with the disposal of sludge, plastic, and glass, mining wastes, drilling fluids, fly and bottom ash, and cement kiln dust (Section 8002), as well as broader studies on resource recovery and conservation (Section 8005(a)). EPA is further directed to collect and evaluate information on waste management and resource recovery, and to maintain a central reference library of such information (Section 8003(a), (b)).

C. Duties of the Interagency Coordinating Committee

The 1980 amendments to the Act established an Interagency Coordinating Committee, chaired by the EPA Administrator or his designee and including representatives of the Department of Energy, Commerce, and the Treasury, and of each other federal agency which EPA determines to have programs or responsibilities affecting resource conservation or recovery (Section 2001(b)). The Committee is responsible for coordinating all federal activities dealing with resource conservation and recovery from solid waste. The Committee is required to submit to Congress annually a 5-year action plan to encourage resource conservation material and energy recovery, and increased private investment in resource conservation or recovery systems. The plan must, at a minimum, describe a coordinated and nonduplicatory approach for federal resource recovery activities.

D. Federal Procurement Guidelines

EPA must prepare procurement guidelines for Federal agencies (Section 6002). After the date fixed in these guidelines, an agency procuring items designated in the guidelines must purchase those items having the highest percentage of recovered materials consistent with maintaining a satisfactory level of competition. The guidelines are to apply only to items whose price exceeds $10,000, or where the quantity of such items purchased in the preceding fiscal year exceeded $10,000 (Section 6002(a)). A decision not to purchase such recycled materials requires a determination that the items (1) were not reasonably available, (2) did not meet performance standards, or (3) were unreasonably priced (Section 6002(c)). By October 1, 1985, all federal agencies responsible for setting procurement specifications must eliminate from those specifications any exclusion of recovered materials. Furthermore, within 1 year after publication of the procurement guidelines or as otherwise specified in those guidelines, such specification must require the use of recovered materials to the "maximum extent possible without jeopardizing the intended end use of the item." (Section 6002(d)).

FOOTNOTES

[1]   *Illinois* v. *Castle,* 12 E.R.C. 1597 (D.D.C. 1979).
[2]   40 C.F.R. Parts 260-265, 122-124; 45 *Fed. Reg.* 33,066-33,588 (May 19, 1980).
[3]   40 C.F.R. § 261.2(b).
[4]   H.R. Rep. No. 94-1491, 94th Cong., 2d Sess. 25 (1976) (hereinafter cited as H. Rep. at -).

5   H. Rep. at 25.

6   40 C.F.R. Part 261.

7   Solid Waste Disposal Act Amendments of 1980, P.L. 96-482.

8   "Hazardous waste generation" is defined as "the act or process of producing hazardous waste". Section 1004(6).

9   The standards are codified at 40 C.F.R. Part 262.

10   The Act itself provides little guidance on the contemplated operation of the manifest system beyond establishing the concept that a special hazardous waste manifest will accompany the waste from the point of generation to the point of ultimate disposal. EPA has implemented this basic legislative concept by establishing a detailed manifest system that tracks hazardous waste from the point of generation to the point of ultimate treatment, storage, and disposal.

11   H. Rep. at 26.

12   A generator may store hazardous wastes on-site for 90 days or less without obtaining a storage permit but even during this 90-day period he must comply with certain minimum storage requirements. (40 C.F.R. § 262.34.)

13   40 C.F.R. §§ 261.5, 261.33.

14   40 C.F.R. §§ 261.5(d), 262.11.

15   40 C.F.R. Part 263.

16   49 C.F.R. Parts 171—179.

17   40 C.F.R. § 260.10(a)(73).

18   Discharges of waste streams subject to Clean Water Act permits are excluded from coverage under this Act. Sections 1004(27), 1006(a). The Act does not contain any provision which exempts wastewater treatment facilities from regulation, however.

19   H.R. Rep. No. 96-1444, 96th Cong. 2d Sess. 33 (1980). Reprinted in (1980) U.S. Code Cong. & Ad. News 8674, 8679.

20   EPA's consolidated permit regulations define an "existing facility" as one which is "in operation" or for which "construction has commenced". 40 C.F.R. § 122.3. The latter phrase includes entering into contractual obligations for construction which cannot be cancelled or modified without substantial loss. *Id.* If a facility was handling a solid waste on or before November 19, 1980 which is subsequently listed or identified as a hazardous waste by EPA, the Agency will consider the facility to have been in existence on that date for interim status purposes. 45 *Fed. Reg.* 76,633 (Nov. 19, 1980).

21   Although EPA cannot grant interim status to facilities which failed to file a timely notification when required to do so, the Agency has announced that it may exercise its enforcement discretion to issue compliance orders allowing such facilities to continue operating if such action is in the public interest, provided they file a permit application and comply with all applicable interim status standards. 45 *Fed. Reg.* 76,632 (Nov. 19, 1980).

22   45 *Fed. Reg.* 76,633 (Nov. 19, 1980).

23   40 C.F.R. Part 265.

24   40 C.F.R. Parts 122 and 124.

25   40 C.F.R. Part 264.

26   The guidelines are codified at 40 C.F.R. Part 123 Subparts A, B, and F.

27   H.R. Rep. No. 96-1444, 96th Cong. 2d Sess. 37 (1980). Reprinted in (1980) U.S. Code Cong. & Ad. News 8674, 8682.

28   *Id.,* at 38. Reprinted in (1980) U.S. Code Cong. & Ad. News 8684.

29   *Id.,* at 37. Reprinted in (1980) U.S. Code Cong. & Ad. News 8683.

30   *United States* v. *Midwest Solvent Recovery, Inc.,* 484 F. Supp. 138, 143-144 (N.D. Inc. 1980).

31   *United States* v. *Solvents Recovery Service,* 14 E.R.C. 2010, 2021 (D. Conn. 1980).

32   *Id.,* at 2022-2023.

33   H. Rep. at 33.

34   Section 4005(a) specifies that schedules in state plans for phasing-out open dumping must require that open dumping practices end by 5 years after the date of publication of the federal criteria on open dumping required by Section 1008(a) of the Act. The criteria were promulgated on September 13, 1979. 45 *Fed. Reg.* (Sept. 13, 1979), codified at 40 C.F.R. Part 257.

35   The guidelines are codified at 40 C.F.R. Part 255.

36   The state plan guidelines are codified at 40 C.F.R. Part 256.

37   The criteria are codified at 40 C.F.R. Part 257.

38   An open dump is any facility or site used for solid waste disposal which does not meet the criteria for a sanitary landfill promulgated under Section 4004. Section 1004(14).

39   The prohibition against new open dumps must take effect no later than the date the state plan is approved. Section 4004(b), (c).

40   During the floor debate, Congressman Rooney explained that the substitute amendment ultimately enacted by the House "[p]ermits States to develop their own schedule for closing open dumps within a 5-year period . . . ." 122 *Cong. Rec.* H 11180 (daily ed., Sept. 27, 1976).

[41] A sanitary landfill is "a disposal site of [sic] which there is no reasonable chance of adverse effects on health and the environment from the disposal of discarded material at the site." H. Rep. at 37. It is statutorily defined as a "facility for the disposal of solid waste which meets the criteria published under section 4004." Section 1004(26). The criteria called for in Section 4004(a) were published on September 13, 1979 and are codified at 40 C.F.R. Part 257.

[42] 45 *Fed. Reg.* 71,818 (Oct. 30, 1980).

[43] 44 *Fed. Reg.* 45,071-72 (July 31, 1979).

[44] 40 C.R.F. § 261.4(a)(2) (Comment). However, EPA has granted a temporary exemption to wastes stored in tanks prior to discharge pursuant to an NPDES permit pending promulgation of special standards and permit requirements for such facilities. 45 *Fed. Reg.* 76,074 (Nov. 17, 1980.

[45] H. Rep. No. 96-144, 96th Cong. 2d Sess. 35 (1980). Reprinted in (1980) U.S. Code Cong. & Ad. News 8674, 8681. The Conferees warned that EPA should be judicious in its use of "private contractors or representatives" for inspection purposes and should avoid the use of contractors who have conflicts of interest.

[46] In *Illinois* v. *Castle,* 12 E.R.C. 1597 (D.D.C. 1979), a suit brought under this provision, the plaintiffs successfully sought judicial enforcement of EPA's nondiscretionary duty to promulgate regulations within the period mandated by statute.

[47] The definition of "recovered materials" excludes materials and byproducts generated from, and commonly reused within, an original manufacturing process. Section 1004(19).

Chapter 6

THE HAZARDOUS WASTE MANAGEMENT PROGRAM UNDER THE RESOURCE CONSERVATION AND RECOVERY ACT*

Cleary, Gottlieb, Steen & Hamilton

TABLE OF CONTENTS

* All footnotes appear at end of chapter.

The U.S. Environmental Protection Agency (EPA) has developed a complicated and far-reaching program to regulate the management of hazardous waste in accordance with the Resource Conservation and Recovery Act (RCRA). The program is designed to track and regulate hazardous wastes from generation to ultimate treatment or disposal, i.e., from "cradle to grave". This is no small task when one considers that in 1981, the first full year that RCRA was in effect, American industry generated 264 million metric tons of hazardous waste.

The hazardous waste program has had a significant impact on industrial operations and long-term planning, and the scope and costs of the hazardous waste program are only going to increase in the decades ahead. The program is now entering its second phase — issuance of hazardous waste permits to treatment, storage, and disposal facilities. Simultaneously, rigorous programs for enforcement of the hazardous waste program are being implemented at both the federal and state levels.[1]

This memorandum first provides an overview of the entire hazardous waste program. Each of the major regulatory areas — standards for generators, transporters, and owners and operators of treatment, storage, and disposal facilities — is then reviewed in more detail. Separate sections review the procedural aspects of interim status and hazardous waste permits, and state hazardous waste programs.

One difficulty encountered in describing the hazardous waste program is the continuous evolution in the regulations. EPA has a number of major rulemakings underway. Litigation — such as that concerning EPA's land disposal regulations — may lead to revisions in the program. Furthermore, another wave of regulation is certain to follow the next round of statutory amendments to RCRA in this or the next Congress. Where possible, this memorandum notes areas of the hazardous waste program which are targets for future revision.

I. OVERVIEW OF THE HAZARDOUS WASTE PROGRAM

A. Identification of Hazardous Waste

EPA's regulations[2] identify the "hazardous waste" subject to RCRA. "Hazardous waste" is a subset of "solid waste" which is broadly defined to include garbage, sludge, and generally any solid, liquid, semi-solid, or contained gaseous material. The Part 261 regulations provide that a waste is "hazardous" either: (1) because it is on one of several lists or (2) it exhibits any of four characteristics — ignitability, corrosivity, reactivity, and "extraction procedure" toxicity.

A waste which is on one of the lists or which exhibits one of the characteristics generally must be managed in accordance with the hazardous waste program. There are, however, a number of exceptions to this general rule. For example, hazardous wastes produced by small quantity generators are exempt from many requirements. A broad exemption also exists for certain hazardous wastes which are recycled or recovered.

B. Standards for Generators

Generators of waste materials are required to evaluate their wastes to determine if they are hazardous under RCRA. Any person who generates a regulated hazardous waste is subject to a variety of reporting, record-keeping, and labeling requirements. A generator is also required to comply with the hazardous waste manifest system that tracks substances shipped off-site for treatment, storage, or disposal. The waste must be properly packaged and labeled before leaving the site of generation. The generator is responsible for determining that the waste subsequently arrives at a hazardous waste facility designated by the generator and authorized to handle that type of waste.

C. Standards for Transporters

A transporter of hazardous wastes must have an EPA identification number and comply with the hazardous waste manifest system for tracking hazardous waste shipments. The transporter must deliver the entire quantity of hazardous waste to the designated treatment, storage, or disposal facility and follow certain record-keeping requirements. A transporter of hazardous wastes also is required to comply with Department of Transportation hazardous materials transportation regulations on labeling, marking, packaging, and placarding which have been incorporated into the RCRA regulations.

D. Requirements for Treatment, Storage, and Disposal Facilities

Facilities that treat, store, or dispose of hazardous wastes (TSD facilities) generally must have either "interim status" or a RCRA permit. Interim status allows facilities that have met certain procedural requirements to treat, store, or dispose of hazardous waste pending issuance of a RCRA permit. During interim status, a facility must comply with the interim status or Phase I standards in Part 265 of the regulations. Once a permit is issued, a facility must comply with the Phase II standards for TSD facilities set out in Part 264 as incorporated into the permit.

The interim status regulations impose relatively few technical or facility design standards. The emphasis of the interim status regulations is on reporting, record-keeping, and maintenance requirements, as well as certain financial responsibility standards. A facility with interim status must also develop written operational plans covering personnel training, contingency and emergency procedures, inspection routines, and facility closure.

Effective November 18, 1980, any facility that treated, stored, or disposed of hazardous waste, and which was not specifically exempted, was eligible for "interim status" for its hazardous waste activities. Generally, in order to have interim status, a facility:

1. Must have been in existence as of November 19, 1980
2. Must have filed a Section 3010 notification of hazardous waste activity by August 18, 1980
3. Must have filed a "Part A" RCRA permit application by November 18, 1980

In addition, EPA has identified certain limited circumstances under which a facility may be granted interim status after November 18, 1980.

Interim status essentially allows a facility to operate pending issuance of a RCRA permit. Interim status may be terminated if a facility fails to comply with applicable hazardous waste regulations, engages in hazardous waste management activities not authorized under the terms of its interim status, or is denied a RCRA permit.

A facility with interim status may either submit a Part B RCRA permit application voluntarily or wait until a request for an application is issued. A facility that does not have interim status is prohibited from operating a treatment, storage, or disposal facility until a RCRA permit is issued.

Development and completion of Part B RCRA permit applications is a costly, time-consuming process. Although EPA is moving slowly in the review and issuance of these permits, significant advance planning is warranted by any company that will eventually need a RCRA permit. A permit application for a new hazardous waste facility must be submitted before construction begins. Over the next several years, most interim status facilities will be "called in" and required to obtain Part B permits.

The standards in Part 264 for permitted TSD facilities generally are more stringent than the interim status standards, particularly for land disposal. Specific design and/or performance standards are set for most types of facilities. In addition, land disposal facilities are required to meet complex ground-water monitoring requirements. All TSD facilities subject to the Part 264 standards must comply with various reporting, record-keeping, inspection, closure, post-closure care, and financial responsibility requirements similar to those for interim status facilities. The Part 264 standards are applied to a facility by incorporation into the "Part B" RCRA permit for that facility.

E. Special Standards

EPA eventually plans to issue standards in Part 266 for specified categories of hazardous waste facilities, including: (1) facilities handling used oil; (2) research facilities; and (3) facilities eligible for class permits. The Agency is presently considering the use of class permits for above-ground, hazardous waste tank storage facilities.

F. State Hazardous Waste Programs

A state can obtain authority to operate a RCRA hazardous waste program in lieu of a program administered by EPA. Approximately 42 states presently have authority to run all or part of the RCRA program in their state. A state seeking such authority must demonstrate that its programs are consistent with, and at least equivalent in stringency to, the EPA hazardous waste program. States may apply for authorization to run their own programs in phases. A state may have authority just for Phase I "interim status" facilities, or it may have authority to run some or all of the components of the Phase II program.

G. Civil and Criminal Penalties

EPA can seek civil penalties of up to $25,000 per day for violations of the hazardous waste regulations. Criminal sanctions are provided for persons who "knowingly" violate certain regulations. EPA has issued regulations covering the procedural rules for the administrative assessment of penalties and the revocation and suspension of permits in the case of violations of RCRA, the hazardous waste regulations, or the terms of a permit. The many states with RCRA programs have comparable penalties and procedures.

II. IDENTIFICATION OF "HAZARDOUS WASTES"

In order to be subject to the RCRA regulations, a company must generate or handle "hazardous wastes". The regulations oblige each generator of waste to determine whether that waste is hazardous. EPA views the accuracy of these determinations as the key to the effectiveness of the hazardous waste program. Accordingly, companies must be able to document the basis for a determination that a particular waste is hazardous or nonhazardous.

A. Checklist for Identifying "Hazardous Wastes"

Subject to certain specific exclusions, the regulations in Part 261 define a "hazardous waste" as any solid waste which meets one of the four identified characteristics of hazardous waste or which appears on one of several lists set forth in the regulations. Any company or person that handles solid wastes is required to determine whether those wastes fall within the definition of hazardous waste. A logical step-by-step approach to this determination is as follows:

1. *Determine whether the substance is a "solid waste" as defined by the regulations.* Since "hazardous wastes" are a subcategory of "solid wastes", a substance is not regulated as a hazardous waste unless it falls within the definition of "solid waste".
2. *Check to see if the regulations specifically exclude the solid waste from classification as a hazardous waste.* Various solid wastes are excluded from being identified as hazardous waste.
3. *For a solid waste which is not excluded, review the hazardous waste lists set out in Subpart D of the Part 261 regulations to determine whether the waste appears on these lists.* If so, it generally must be managed as a hazardous waste.
4. *For a solid waste which is not specifically listed as hazardous, determine whether it meets one of the characteristics of hazardous waste.* The four characteristics set out in Subpart C of the Part 261 regulations are ignitability, reactivity, corrosivity, and extraction procedure toxicity.
5. *Determine whether a waste determined to be hazardous is subject to any special procedures or requirements.* Part 261 of the regulations sets out special provisions for hazardous waste which is (a) used, reused, recycled, or reclaimed; (b) generated in small quantities of hazardous wastes; (c) a residue in "empty" containers; (d) generated in product or raw material storage tanks, transport vehi-

cles, and pipelines, or in manufacturing units, if it has not exited the unit in which it was generated; or (e) generated in waste-water treatment units or primary neutralization tanks.

If a waste falls within one of these categories, it is exempt from some or all of the requirements of the hazardous waste program. See Section II.F *infra*.

B. The Definition of "Solid Waste"

Solid wastes fall into two categories: (1) garbage, refuse, or sludge; and (2) certain waste material which is "solid, liquid, semi-solid, or contained gaseous material, resulting from industrial, commercial, mining, or agricultural operations, or from community activities" (§ 261.2(a), (b)). *All* garbage, refuse, and sludge is solid waste. A material in the second category is a solid waste if it (1) has been discarded or is being accumulated, stored, or physically, chemically, or biologically treated prior to being discarded; *or* (2) has served its intended purpose and *is sometimes discarded; or* (3) is a manufacturing or mining by-product and *is sometimes discarded* (§ 261.2(b)).[3]

The inclusion of material which has served its intended purpose and is sometimes discarded is meant to cover materials such as waste solvents, paint wastes, waste acids, and waste oil.[4] EPA also includes within that term solid waste materials which are incidentally generated during mining and manufacturing operations. The regulations exclude primary products and intermediate products (§ 261.2(e)).

The regulations (§ 261.4(a)) specifically exclude the following materials from the definition of solid waste:

1. Domestic sewage, including domestic sewage which is mixed with industrial waste-waters in a public sewer treatment system
2. Industrial waste-water discharges from point sources[5]
3. Nuclear or nuclear by-product material covered by the Atomic Energy Act of 1954
4. Irrigation return flows
5. Materials subjected to *in situ* techniques which are not removed from the ground as part of the extraction process

C. Solid Wastes which are Not Hazardous Wastes

The regulations specify that the following solid wastes are not hazardous:

1. Household waste, which is defined as "any waste material (including garbage, trash, and sanitary wastes in septic tanks)" derived from households (including single and multiple residences, hotels and motels)"
2. Solid wastes which are returned to the soil as fertilizers *and* are generated by either the growing or harvesting of agricultrual crops or the raising of animals (including animal manures)
3. Mining overburden which is returned to the mine site
4. Fly ash waste, bottom ash waste, slag waste, and flue gas emission control waste generated primarily from the combustion of coal or other fossil fuels
5. Drilling fluids and other wastes associated with the exploration, development, or production of crude oil, natural gas, or geothermal energy

D. The Hazardous Waste Lists

Subpart D of the Part 261 regulations contains four separate lists of hazardous wastes: (1) wastes from processes used in a variety of industries (§ 261.31); (2) wastes from processes used in particular segments of industry (§ 261.32); (3) acutely hazardous materials (§ 261.33(e)); and (4) toxic materials (§ 261.33 (f)). Each listed waste is assigned an EPA identification number to be used in complying with various reporting and record-keeping requirements. EPA continues to add new wastes to these lists.

1. Lists of Hazardous Wastes and Waste Processes

The first two hazardous waste lists are entitled Hazardous Waste from Nonspecific Sources (§ 261.31) and Hazardous Waste from Specific Sources (§ 261.32). The items on the list from nonspecific sources include spent solvents, wastes from electroplating and heat treating, and other "generic" wastes which may result from manufacturing operations in a number of industries. The wastes on the list from specific sources include particular wastes from: inorganic pigment and chemical production; wood preservation; organic chemical and pesticide manufacturing; petroleum refining; leather tanning; and ferrous and nonferrous metal production. Any solid waste containing a waste on these two lists must be managed as a hazardous waste unless otherwise exempted by the regulations.[7]

2. Lists of Acutely Hazardous and Toxic Products and Related Materials

The third and fourth hazardous waste lists are entitled "Discarded Commercial Chemical Products, Off-Specification Species, Container Residues, and Spill Residues Thereof" (§ 261.33(e), (f)). Both lists identify certain commercial products and materials which become hazardous wastes when they are discarded or intended to be discarded.[8]

The third list (§ 261.33(e)) contains almost 200 materials and products which EPA has found to meet the criteria for *acutely hazardous* wastes (§ 261.11(a)(2)). Examples of acutely hazardous wastes are acrolein, carbon disulfide, cyanides, and a large number of pesticides.

The fourth list (§ 261.33(f)) consists of over 450 materials and products listed on the basis of the criteria for *toxicity*. This list includes substances such as acetone, creosote, methanol, phenol, and toluene.

It is important to note that both lists are limited to chemical substances manufactured or formulated for commercial or manufacturing use or their off-specification species.[9] These substances only become hazardous wastes if and when the listed chemical product itself is discarded or intended to be discarded. For example, such materials might be discarded due to failure to meet required specifications, inventory reduction, or changes in the product line. Thus, the mere presence in a waste, such as a process waste-water or emission control device purge stream, of a substance identified on these lists does not result in classification of the waste as hazardous if the substance is not a discarded commercial chemical product.[10]

The hazardous waste program also regulates the disposal of debris from spills or other discharges[11] of materials on either of the commercial products lists (§ 261.33(d)). Any "residue or contaminated soil, water or other debris resulting from cleanup of a spill" of a listed substance is itself a hazardous waste. However, if the cleanup from a discharge of an *acutely hazardous* waste (the § 261.33(e) list) is less than 100 kg, or if a discharge cleanup results in less than 1000 kg of waste from a discharge of a *toxic* waste (the § 261.33(f) list), then the debris may be excluded from many requirements of the hazardous waste program under the small generator exemption (§§ 261.5(a), (e)(2)). Therefore, subject to the quantity exclusions, debris from the cleanup of soil containing the discharged compound is subject to the regulations regardless of the quantity of the compound in the debris.

E. The Characteristics of Hazardous Waste

Subpart C of Part 261 sets out the characteristics of hazardous waste. A solid waste which is not specifically listed as a hazardous waste may still be hazardous if it meets one of the four characteristics of hazardous waste — i.e., ignitability, reactivity, corrosivity, and extraction procedure toxicity (§§ 261.21-.24). Hazardous wastes which exhibit one of these characteristics but are not on the hazardous waste lists in Subpart D are assigned the EPA Hazardous Waste Number which corresponds to the appropriate characteristic. This identification number is to be used when complying with the various reporting and record-keeping requirements.

1. The Ignitability Characteristic

If a "representative sample"[12] of a solid waste exhibits any of the following properties, then the solid waste is a hazardous waste on the basis of ignitability:

1. It is a liquid, other than an aqueous solution containing less than 24% alcohol by volume, and has a flash point less than 60°C (140°F);
2. It is not a liquid and is capable, under standard temperature and pressure, of causing fire through friction, absorption of moisture, or spontaneous chemical changes and, when ignited, burns so vigorously and persistently that it creates a hazard;
3. It is an ignitable compressed gas as defined in 49 C.F.R. § 173.300 of the Hazardous Materials Regulations of the Department of Transportation; or
4. It is an oxidizer as defined in 49 C.F.R. § 173.151 of the Hazardous Materials Regulations (§ 126.21).

In defining the characteristics of ignitability, EPA sought to "identify wastes capable of causing fires during routine transportation, storage, and disposal and wastes capable of severely exacerbating a fire once started". 45 *Fed Reg.* 33108.

2. The Corrosivity Characteristic

If a representative sample of a solid waste exhibits any of the following properties, then the solid waste is a hazardous waste based on corrosivity:

1. It is aqueous and has a pH greater than or equal to 2 or less than or equal to 12.5; or
2. It is a liquid and corrodes steel (SAE 1020) at a rate greater than 6.35 mm (0.250 in.) per year at a test temperature of 55°C (130°F) (§ 261.22(1)).

3. The Reactivity Characteristic

If a representative sample of a solid waste has any of the following properties, then the waste is hazardous because it meets the characteristic of reactivity:

1. It is normally unstable and readily undergoes violent change without detonating;
2. It reacts violently with water;
3. It forms potentially explosive mixtures with water;
4. When mixed with water, it generates toxic gases, vapors, or fumes in a quantity sufficient to present a danger to human health or the environment;
5. It is a cyanide- or sulfide-bearing waste which, when exposed to pH conditions between 2 and 12.5, can generate toxic gases, vapors, or fumes in a quantity sufficient to present a danger to human health or the environment;
6. It is capable of detonation or explosive reaction if it is subjected to a strong initiating source or if heated under confinement;
7. It is readily capable of detonation or explosive decomposition or reaction at standard temperature and pressure; or

8.   It is a forbidden explosive as defined in 49 C.F.R. § 173.51 or a Class A explosive as defined in 49 C.F.R. § 173.53 or a Class B explosive as defined in 49 C.F.R. § 173.88. (§ 261.23(a).)

EPA was unable to find specific, quantifiable properties of reactive waste which could be measured by standardized test protocols. However, EPA takes the position that, despite the lack of specific tests, most generators of reactive wastes are aware that their wastes possess this characteristic and can handle them accordingly. 45 *Fed. Reg.* 33110.

4. The Extraction Procedure Toxicity Characteristic

The determination whether a waste meets the Extraction Procedure (EP) toxicity characteristic requires use of the Extraction Procedure test in Appendix II to Part 261 or an equivalent test approved by EPA. The extract from the Extraction Procedure must be analyzed for the presence of 14 contaminants for which there are National Interim Primary Drinking Water Standards. If the extract[13] contains any of the contaminants at a concentration equal to or greater than the respective value given below, then the waste is hazardous on the basis of EP toxicity (§ 261.24(a)):

Maximum Concentration of Contaminants
for Characteristic of EP Toxicity

Maximum concentration (mg/l)	Contaminant
5.0	Arsenic
100.0	Barium
1.0	Cadmium
5.0	Chromium
5.0	Lead
0.2	Mercury
1.0	Selenium
5.0	Silver
0.02	Endrin
0.4	Lindane
10.0	Methoxychlor
0.5	Toxaphene
10.0	2,4-D,
1.0	2,4,5-TP Silvex

These maximum concentration levels for each contaminant are 100 times the National Interim Primary Drinking Water Standards.[14]

F. Special Provisions for Certain Categories of Hazardous Wastes

Certain categories of hazardous waste are not subject to all of the hazardous waste provisions or are not subject to those requirements immediately upon generation. These categories are: (1) hazardous wastes which are used, reused, recycled, or reclaimed; (2) hazardous waste produced in small quantities; (3) wastes from a particular facility which have been delisted by a rulemaking petition; (4) residues of hazardous wastes in empty containers; (5) hazardous wastes generated in manufacturing process units, storage tanks, transport vehicles or vessels, or pipelines; and (6) samples of hazardous wastes.

1. Hazardous Wastes which are Used, Reused, Recycled, or Reclaimed

One of the more controversial aspects of the hazardous waste program is regulation of materials to be recycled or reused. Companies that recycle or reuse hazardous wastes should watch for new regulatory developments in this area.[15]

Under the current regulations, materials which are used, reused, recycled, or reclaimed are "solid wastes" and therefore subject to regulation as hazardous waste if they are discarded or "sometimes discarded" (§ 261.2(b)). Therefore, a material could be considered a "waste" even though its generator has never discarded it in the past and has no intent to discard it in the future. EPA justified this approach by arguing that otherwise a broad loophole would exist because the determination whether a waste is subject to the regulations would depend on the intent or practice of the person handling it. 45 *Fed. Reg.* 33091.

Companies that reclaim or recover waste material should realize that EPA's broad definition of the term "treatment" also includes recovery operations. The regulations define "treatment" as:

> "[A]ny method, technique or process, including neutralization, designed to change the physical, chemical, or biological character or composition of any hazardous waste so as to neutralize such waste, or *so as to recover energy or material resources from the waste,* or so as to render such waste non-hazardous, or less hazardous; safer to transport, store or dispose of; or amenable reduced in volume" (§ 260.10). (Emphasis added.)

The italicized language was added in the final regulations and is not part of the statutory definition of "treatment" set out in RCRA.

While EPA initially gathered all of these "wastes" which are "sometimes discarded" within its regulatory net, it did not issue comprehensive standards regulating the use, reuse, recycling, and reclamation of hazardous wastes. Accordingly, material which is used, reused, or reclaimed generally is presently *exempt* from the hazardous waste program *if* it meets any of the following criteria:

1. It is being beneficially used or reused or legitimately recycled or reclaimed;
2. It is being accumulated, stored, or physically, chemically, or biologically treated prior to beneficial use or reuse or legitimate recycling or reclamation; or
3. It is spent pickle liquor which is reused in waste-water treatment at an NPDES-permitted facility, or treated before such reuse (§ 261.6(a)).

EPA has emphasized that sham recycling operations, such as burning organic wastes with little or no heat value in industrial boilers, are not within the scope of this exemption and will be subject to enforcement actions under Section 3008 of RCRA. 45 *Fed. Reg.* 33093. In addition, the Agency intends to develop regulations controlling the burning of hazardous waste in boilers.[16]

The current regulations do place restrictions on *transportation* and *storage* of certain materials which are used, reused, recycled, or reclaimed. Any hazardous waste which: (1) is a sludge; (2) is listed as hazardous in § 261.31 or § 261.32; *or* (3) contains a waste listed in § 261.31 or § 261.32; *and* which is transported or stored prior to being used, reused, recycled, or reclaimed (§ 261.6(b)) is subject to the following requirements with respect to such transportation or storage:

1. The notification requirement of Section 3010 of RCRA
2. The standards in Part 262 for generators of hazardous waste
3. The standards in Part 263 for transporters of hazardous waste
4. Most of the standards for permitted treatment, storage, and disposal facilities (Subparts A—L of Part 264)
5. Most of the interim status standards for treatment, storage, and disposal facilities (Subparts A—L of Part 265)

6.   The permit requirements which apply to hazardous waste storage facilities (Parts 270 and 124) (§ 261.6(b))

2. Hazardous Wastes Produced by Small Quantity Generators

The current regulations[17] contain certain exemptions for hazardous wastes produced by "small quantity generators". The exemption is structured in a two-tier fashion: it sets both a general exemption level and significantly lower exemption levels for "acutely hazardous" wastes and related material.

A generator that generates a total of less than 1000 kg of hazardous wastes at a particular site in a calendar month is a "small quantity generator", and its waste is generally exempt from the hazardous waste regulations (§ 261.5(a) and (b)). However, if more than 1000 kg of such wastes are accumulated, the small generator exemption is no longer applicable (§ 261.5(f)). Once the exemption level is exceeded, all of those wastes and any wastes subsequently added to that accumulation are subject to regulation. However, a generator can accumulate waste in excess of 1000 kg in on-site storage for up to 90 days without having a RCRA permit or interim status.[18] Waste which was counted toward the 1000 kg exemption level when generated is not counted again toward that level when removed from on-site accumulation or storage (§ 261.5(d)). This 1000-kg/month exemption applies to *all* hazardous wastes *except* those which are listed as acutely hazardous.

The regulations specify lower exemption levels for acutely hazardous wastes listed in § 261.33(e):

Acutely hazardous waste	Exemption level
Commercial product or chemical intermediate having the generic name on the acutely hazardous waste list, or an off-specification variation of such a product or intermediate (§§ 261.33(a,)(b))	1 kg or less per month (§ 261.5(e)(1))
Residues in nonempty containers of acutely hazardous wastes (§ 261.33(c))	1 kg or less per month (§ 261.5(e)(1))
Residue, contaminated soil or other debris from a spill of an acutely hazardous waste (§ 261.33(d))	100 kg/month or less (§ 261.5(e) (2))

If a generator generates in 1 month or accumulates at any time acutely hazardous wastes in quantities greater than the levels set out above, or if he generates acutely hazardous wastes at lower levels but produces more than 1000 kg of hazardous waste a month, all of his acutely hazardous waste is subject to all applicable requirements of the hazardous waste program (§ 261.5(e)).

At both the 1000-kg level and the more stringent levels for acutely hazardous wastes, the availability of the small quantity exemption is dependent upon compliance with certain conditions (§ 261.5(g)). First, an exempted small quantity generator — like all generators — is still required to make an initial determination of whether its waste is hazardous (§ 261.11). Second, the generator must either treat, reuse, or store the waste on-site or ensure its delivery to: (1) a hazardous waste facility with either interim status or a RCRA permit; (2) a facility permitted, licensed, or registered by a state to manage municipal or industrial solid waste; or (3) a facility which beneficially reuses or recycles the waste, or treats the waste prior to such recycling or reuse (§ 261.5(g)).

Hazardous waste which is exempt under the small quantity generator exemptions may be mixed with any quantity of nonhazardous waste and still remain exempt so long as the resulting mixture does not meet any of the four hazardous waste characteristics.

3. Exclusion of a Particular Facility's Waste by Rule-Making Petition

A generator may petition EPA to exclude from the hazardous waste lists a specific waste generated at a particular facility (§ 260.22). Generally, a petitioner must demonstrate to the satisfaction of EPA that the waste in question does not meet any of the criteria for which the waste or its constituents were listed as hazardous and that it does not meet any of the hazardous waste characteristics. The regulations specify the information to be included in a petition to exclude waste from a particular facility (§ 262.22(i)) and provide that EPA may request additional information (§ 260.22(j)). If the waste in question is a mixture of nonhazardous waste and one or more listed hazardous wastes, or is derived from one or more listed hazardous wastes, the demonstration may be made with respect to each constituent listed waste or the waste mixture as a whole (§ 260.22(b)).

If a waste or any of its constituents is listed as an acutely hazardous waste, then the petitioner must show that the waste does not meet the criteria for listing as either an acutely hazardous waste or a toxic waste (§ 260.22(e)). In other words, it is necessary to show not only that the waste no longer is extremely toxic but also that, for those wastes which contain a constituent listed in Appendix VIII to Part 261, consideration of the enumerated factors in Section 261.11(a)(3) indicates that the waste is not capable of posing a substantial hazard to health or the environment when improperly managed. It is not sufficient to demonstrate that the waste is not hazardous because it is being properly managed by the petitioner.

To exclude wastes or waste constituents that have been listed as ignitable, reactive, corrosive, or Extraction Procedure toxic, the petitioner must demonstrate that the wastes or constituents do not exhibit the relevant characteristics as defined in Part 261 (§ 260.22(c)). To exclude wastes or constituents listed as toxic, the petitioner must demonstrate either that: (1) a sample of the waste or constituent does not contain the particular substance that caused EPA to list the waste as toxic;[19] or (2) even though the waste contains one of the substances which caused EPA to consider listing a waste as toxic, consideration of the factors enumerated in Section 261.11(a)(3) indicates that the waste is not capable of posing a substantial hazard to health or the environment when improperly managed (§ 260.22(d)).

4. Hazardous Waste Residues in "Empty" Containers

Hazardous wastes are often stored or transported in containers. The typical methods of emptying such containers (i.e., pouring, pumping, or aspirating) are not capable of removing all residues. Accordingly, EPA has developed its own definition of "empty". Hazardous wastes remaining in an "empty" container or in an inner liner removed from an "empty" container are not subject to the hazardous waste regulations. Hazardous waste residues in containers that do not fit EPA's definition of "empty", or the inner liners of such containers, *are* subject to the regulations (§ 261.7(a)).

The definition of "empty" container varies depending on the type of hazardous waste in the container. The regulations identify three categories of containers: (1) containers that have held acutely hazardous wastes; (2) containers that have held compressed gas; and (3) containers that have held other types of hazardous waste.

Residues of acutely hazardous materials in containers, or inner liners from such containers, which are discarded or intended to be discarded are subject to regulation unless the container meets the following definition of "empty" (§ 261.33(c)):

1.   The container or inner liner has been triple rinsed with a solvent capable of removing the listed products;
2.   The container or liner has been cleaned by another method demonstrated to achieve equivalent removal; or
3.   In the case of a container, the liner that prevented contact between the container and the acutely hazardous material has been removed (§ 261.7(b)(3)).

A container that has held a compressed gas that is a hazardous waste is considered empty when the pressure in the container approaches atmospheric (§ 261.7(b)(2)).

Residues of hazardous wastes which are not acutely hazardous or gaseous in containers or inner liners which are discarded or intended to be discarded are regulated unless:

1.   All wastes have been removed that can be removed using the practices commonly employed to remove materials from that type of container; *and*
2.   No more than 2.5 cm (1 in.) of residue remains on the bottom of the container or inner line.[20]
3.   No more than 3% by weight of the total capacity of the container remains in the container or inner liner if the container is less than or equal to 110 gal in size. If the container is greater than 110 gallons, no more than 0.3% can remain. (§ 261.7(b)(1)).[21]

If the hazardous waste is a two-phase mixture of a liquid and a nonviscous solid or semi-solid, the container cannot be considered empty if the waste adhering to the sides of the container would amount to more than a 1-in. layer at the bottom. 45 *Fed. Reg.* 78525-26 (Nov. 25, 1980).

5. Hazardous Wastes Generated in Manufacturing Units and Product or Raw Material Storage Tanks, Transport Vehicles, or Pipelines

Hazardous waste that is generated in manufacturing units, product or raw material storage tanks, transport vehicles, or pipelines generally is not subject to the hazardous waste regulations until: (1) it exits the unit in which it was generated; or (2) the waste has remained in the unit for more than 90 days after the unit ceases to be operated for manufacturing or for transporting product or raw materials (§ 261.4(c)). This exemption does not apply to wastes that are generated or accumulated in surface impoundments.

Hazardous wastes that remain in product or raw material pipelines or storage tanks for more than 90 days after shutdown are subject to full regulation under the hazardous waste management program. The pipeline, storage tank, or other unit containing such waste would be deemed a hazardous waste storage facility for which interim status or a RCRA permit would be required. If the cessation of operations is permanent and the owner or operator intends the pipeline or tank to be the ultimate repository of the hazardous waste sediment or residue, it is considered a disposal facility for regulatory purposes. See 46 *Fed. Reg.* 2807 (Jan. 12, 1981).

EPA recognizes that frequently one person owns and operates a storage tank or transport vehicle, a second owns the product or raw material being stored or transported, and a third performs the actual removal of the hazardous waste sludges, sediments, and residues from these units. The Agency considers all three parties to be potentially jointly and severally liable as "generators".[22] EPA has indicated, however, that it would look initially to the operator of the storage tank or transport vehicle to perform the generator responsibilities. 45 *Fed. Reg.* 72026-27 (October 30, 1980). Accordingly, users of storage and pipeline facilities owned by others may want to consider contractual provisions that require the facility owner to comply with all applicable RCRA regulations and to hold the user harmless for any failure to do so.

6. Samples of Hazardous Waste

If a waste sample is handled in accordance with certain conditions, the sample is not subject to the hazardous waste regulations during:

1. Transportation to and from a laboratory used to test the waste sample;
2. Storage of the sample before transport to the laboratory;
3. Storage of the sample at the laboratory before or after testing, prior to return to the sample collector; or
4. Storage of the sample at the laboratory for a specific purpose (such as holding as evidence)

The exemption from the regulations only applies so long as the sample is handled in accordance with Department of Transportation and other applicable shipping requirements, is securely packaged, and is accompanied by certain minimum identifying information spelled out in the regulations (§ 261.4(d)).

G. Events Subsequent to Generation which can Affect the Determination Whether a Waste is Hazardous

Under certain circumstances, the handling of a hazardous waste after generation may change its classification as a hazardous waste. For example, a hazardous waste may be mixed with nonhazardous wastes or treated or disposed of in a manner that results in the composite waste no longer meeting the definition of hazardous waste. The regulations address these situations in some detail.

As a general rule, if a listed hazardous waste is mixed with *any* quantity of nonhazardous waste, the resulting mixture is still a hazardous waste (§ 261.3(a)(2)). However, EPA has created a number of exceptions to this "mixture rule" for listed hazardous wastes. Waste-water subject to regulation under Sections 402 or 307(b) of the Clean Water Act is not considered a hazardous waste under the mixture rule if the wastewater would only be hazardous because:

1. It contains certain spent solvents at no more than specified concentrations;
2. It contains heat exchanger bundle sludge from the petroleum refining industry;
3. It contains a discarded commercial product listed in § 261.33 in amounts due solely to "de minimis" losses of these materials during normal manufacturing operations; or
4. It contains less than specified quantities of wastes from laboratory operations that are listed as hazardous due to toxicity (T) (§ 261.3(a)(2)(iv)).

The mixture rule does not apply to mixtures that contain wastes listed as hazardous due to one of the four characteristics if the mixture itself no longer exhibits any of the four characteristics (§ 261.3(a)(2)(iii)). A mixture of nonhazardous wastes and wastes which are not listed but which meet the characteristics of hazardous wastes also is hazardous only if the mixture exhibits one of the hazardous waste characteristics (§ 261.3(b)(3)). Thus, with respect to wastes which are hazardous only because they meet one of the four characteristics, the burden of compliance with the hazardous waste program might be reduced by mixing these wastes so that they are no longer hazardous. However, the mixing of hazardous wastes with other wastes in order to render the mixture nonhazardous or less hazardous qualifies as "treatment" (§ 260.10). Treatment of hazardous waste usually requires interim status or a RCRA permit.

The regulations also provide that "any solid waste generated from the treatment, storage, or disposal of a hazardous waste" is a hazardous waste, including any sludge, spill residue, ash, emission control dust, or leachate (§ 261.3(c)(2)).

III. STANDARDS FOR HAZARDOUS WASTE GENERATORS

The standards for generators in Part 262 apply to "any person, by site, whose act or process produces hazardous waste identified or listed in Part 261 . . . or whose act first causes a hazardous waste to become subject to regulation" (§ 260.10). The first obligation of any generator is to determine whether the wastes produced by that generator are "hazardous wastes". If so, the generator must comply with the requirements in Part 262 regarding reporting and record-keeping, use of the manifest system to track hazardous waste shipments, and proper packaging and labeling of hazardous wastes prior to transport. A generator is also responsible for selecting an appropriate facility to treat, store, or dispose of its wastes. Generators engaged in on-site hazardous waste treatment, storage, or disposal operations must have interim status and comply with the applicable interim status regulations.

A. Obligation to Determine if a Solid Waste is Hazardous or Nonhazardous

Any person who generates solid waste is required to make an evaluation of whether the waste is hazardous or nonhazardous under Part 261 of the regulations (§ 262.11). The regulations do not allow the generator the option of simply declaring waste to be hazardous and then managing it as such. 45 *Fed. Reg.* 12727 (Feb. 26, 1980).

The regulations make no provision for a good faith error in characterizing waste as hazardous or not. Furthermore, generators must maintain records "of any test results, waste analyses, or other determination made" in evaluating the hazardousness of a solid waste (§ 262.40(c)). These records should be reviewed carefully to ensure that they adequately substantiate each determination. The retention period for such records is at least 3 years from the date the waste was last sent to on-site or off-site treatment, storage, or disposal. The retention period is automatically extended for the duration of an unresolved enforcement action or upon request by EPA (§ 262.40(d)).

In determining whether a waste is hazardous, a generator should be certain to review each of the four hazardous waste lists (§§ 261.31, 261.32, 261.33(e) and 261.33(f)). A generator who produces any listed wastes should also consider whether it generates any of the following hazardous wastes:

1. A mixture which contains any quantity of a listed hazardous waste (regardless of the amount of nonhazardous waste in the mixture) and does not fall within one of the exemptions to the mixture rule (§ 261.3(a)(2)(iv);
2. A waste derived from a waste listed in Subpart D, i.e., waste generated from the treatment, storage, or disposal of a waste listed in Subpart D (§ 261.3(c)(2); or
3. Any residue, contaminated soil or water or other debris from cleanup of a spill of a waste listed as acutely hazardous (§ 261.33(3)) or toxic (§ 261.33(f)) (see § 261.33(d))

If a waste is not listed, the generator must determine whether the waste exhibits any of the four characteristics of hazardous waste discussed previously: ignitability, corrosivity, reactivity, and Extraction Procedure toxicity. This determination may be made either: (1) by testing the waste for these characteristics (§ 262.11(c)(1)),[23] or (2) by assessing the probable hazardous characteristics of the waste in light of the materials and process used in generating the waste (§ 262.11(c)(2)).

A generator of a listed hazardous waste or a waste that exhibits one of the characteristics should then review the various exemptions in Part 261 and the small generator exemption to see if an exemption is applicable.

B. Frequency of Determinations

No specific frequency is established for the testing or retesting of a generator's wastes. However, a generator has a continuing responsibility to determine whether the wastes are hazardous. EPA expects manufacturers to be familiar with any "significant" changes in their materials, processes, or operations, and to determine whether these changes have any impact on the hazardousness of the wastes generated. If testing is necessary in order to be certain of the hazardousness of the wastes, then the generator is obligated to conduct such tests.

C. EPA Identification Numbers

A generator may not treat, store, dispose of, transport, or offer for transportation hazardous waste without receiving an EPA Identification Number. A generator similarly cannot offer hazardous waste to treatment, storage, or disposal facilities that have not received EPA Identification Numbers (§ 262.12). Generators may obtain an EPA Identification Number by submitting a notification form (EPA Form 8700-12) to EPA.

D. Storage of Hazardous Wastes on the Site of Generation

Almost all generators store hazardous waste on the site of generation for some time, whether prior to transportation for off-site treatment, storage, or disposal or prior to on-site treatment or disposal. Storage of hazardous waste generally requires either interim status or a RCRA permit. However, EPA has exempted certain short-term, on-site storage by generators from this permit requirement.

A generator may store hazardous wastes on-site[24] for 90 days or less without either a permit or interim status *if* the following conditions are met:

1. The hazardous waste is stored in tanks and containers which comply with the interim status standards in Part 265, except § 265.193;
2. The date upon which accumulation begins is clearly marked on each container;
3. While the waste is being accumulated on-site, each container and tank is clearly labeled "Hazardous Waste", and
4. The generator complies with the emergency preparedness and prevention, personnel training, and contingency plan requirements set out in Part 265 for facilities with interim status (§ 262.34(a)).

These requirements apply to both initial satellite collection points and centralized accumulation facilities.[25]

A generator who stores hazardous waste for more than 90 days generally is considered an operator of a storage facility requiring interim status or a permit. However, an extension to the 90-day period *may* be granted by EPA if hazardous wastes must remain on-site for longer than 90 days due to "unforeseen, temporary, and uncontrollable circumstances" (§ 262.34(b)).

E. On-Site Treatment and Disposal of Hazardous Wastes

Any generator who treats or disposes of hazardous wastes on-site is considered an owner or operator of a hazardous waste treatment or disposal facility requiring either interim status or a permit (§ 262.10 (Note)). As an owner or operator, the generator must then also comply with the applicable requirements for such facilities set out in Parts 264 and 265.

Generators should periodically review operations to check if any activities qualify as "treatment" or "disposal", as these terms are broadly defined in the RCRA regulations.[26] For example, the definition of "treatment" encompasses the mixing of wastes to render them less hazardous, adjustment of pH, and many other processes which generators might not view as waste treatment.

F. Use of the Manifest System to Track Hazardous Waste Shipments

The hazardous waste program establishes a manifest system to track hazardous waste shipments off the site of generation to ultimate treatment, storage, or disposal. Generators are responsible for selecting a particular TSD facility to receive their wastes and are responsible for determining that the facility has the necessary permits or authorization for handling a particular type of waste (§ 262.20(b)).

In order to determine whether a facility has received interim status or a permit, a generator should ask the facility for its EPA Identification Number and for a certification that it has been given interim status or a permit. If the generator has any doubt about the status of a facility, the generator should check with the appropriate EPA Regional Office prior to sending waste to that facility.

The generator has the option of designating on the manifest an alternate hazardous waste facility to which the particular waste shipment can be delivered in case an emergency prevents delivery to the primary designated facility (§ 262.20(c)). If the transporter is unable to deliver the shipment to either the original or alternate (if any) designated facility, the generator must instruct the transporter either to take the waste to another facility designated by the generator or to return the waste shipment to the generator (§ 262.20(d)).

Each generator who transports hazardous waste or offers it for transportation to an off-site facility must complete a hazardous waste manifest prior to transportation off-site.[27] Effective September 20, 1984, companies must use the uniform manifest form issued in a joint rulemaking by EPA and DOT. (See 40 C.F.R. § 262.20(a), as amended at 49 *Fed. Reg.* 10490 (March 20, 1984).) Until that date, companies must be certain that whatever shipping form is used contains all the information required by the regulations (§ 262.21(a)).

Before a hazardous waste shipment leaves the site of generation, the generator must sign the manifest certification, obtain the transporter's signature on the manifest, and note the date of acceptance of the shipment (§ 262.23(a)). The generator is to retain one copy of the manifest and generally must provide the transporter with enough copies for each transporter and two copies for the designated facility.[28]

Generators are also responsible for ensuring that hazardous waste shipments actually arrive at and are accepted by the designated hazardous waste facilities. A designated facility is required to sign and return a copy of the manifest to the generator within 30 days after receipt of a hazardous waste shipment. If the generator has not received a signed copy of a manifest from a facility within 35 days after acceptance of the waste by the transporter, the generator must contact the transporter and/or the designated facility to determine the status of the waste shipment (§ 262.42(a)).

If the generator has not received a signed copy of the manifest from a facility within 45 days after acceptance of the waste by the transporter, the generator must submit an Exception Report to the EPA Regional Administrator for the Region in which the generator is located (§ 262.42(b)). The Exception Report must include a legible copy of the manifest for the shipment and a cover letter explaining the generator's efforts to locate the hazardous waste shipment and the results of those efforts.[29]

G. Preparation of Hazardous Wastes for Transportation Off the Site of Generation

Before transporting or offering hazardous wastes for transportation, a generator must comply with applicable Department of Transportation (DOT) regulations for packaging, labeling, marking, and placarding hazardous materials. EPA has coordinated its regulations and enforcement program regarding the transportation of hazardous wastes with DOT's existing program for the transportation of hazardous materials.

Generators are required to package hazardous wastes in appropriate tanks and containers in accordance with the DOT regulations in 49 C.F.R. Parts 173, 178, and 179 (§ 262.30). Generators must mark, label, and placard hazardous wastes in accordance with the DOT requirements set out in 49 C.F.R. Part 172 (§§ 262.31-.33).

H. Reporting and Record-Keeping Requirements

Part 262 imposes the following record-keeping requirements on generators of hazardous wastes (§ 262.40):

Records	Retention time
A copy of each signed manifest returned by a treatment, storage, or disposal facility, and copies of any manifests signed by the generator and transporter for which no signed manifest has been returned from a treatment, storage, or disposal facility (§ 262.40(a))	3 years from the date of acceptance of the shipment by the initial transporter
A copy of each Biennial Report and Exception Report filed by the generator (§ 262.40(b))	3 years from the due date of the report
Records of any test results, waste analyses, or other determinations made by a generator in determining whether a solid waste is a hazardous waste under Part 261 (§ 262.40(c))	3 years from the date that the waste was last sent to on-site or off-site treatment, storage, or disposal

The retention periods set out above are automatically extended during an enforcement action regarding the regulated activity or as requested by EPA (§ 262.40(d)).

A generator who ships hazardous waste off-site must submit a Generator Biennial Report using EPA Form 8700-13A. The report must be sent to the Regional Administrator for the Region in which the generator is located by March 1 of each even numbered year for the preceding calendar year (§ 262.41(a)). The report requires detailed information on *each* shipment of hazardous waste off-site during the preceding year.

Any generator who treats, stores, or disposes of hazardous waste on-site must submit a Facility Biennial Report covering those wastes (§ 262.41(b)). Depending on whether the generator has a permit or interim status, it must comply with the requirements for Biennial Reports set out in Parts 264 or 265. Generally, the generator must file a report by March 1 of each even numbered year for the preceding calendar year. The report requires information on the method of treatment, storage, or disposal of each hazardous waste, monitoring data, and estimates of closure and post-closure costs.

IV. STANDARDS FOR SHIPPERS AND TRANSPORTERS OF HAZARDOUS WASTE

The Part 263 standards for transporters of hazardous waste apply whenever the use of a hazardous waste manifest is required by the Part 262 regulations.[30] The Part 263 regulations require transporter identification numbers, compliance with the manifest system and certain DOT hazardous materials regulations,[31] and mandate reporting and cleanup in the event of a discharge during transportation.

Transporters are cautioned that they become subject to the regulations applicable to *generators* (Part 262) if they transport hazardous waste into the U.S. from abroad or if they mix wastes with different DOT shipping descriptions by placing them in the same container (§ 263.10(c)). In addition, transporters who hold manifested shipments of hazardous waste at a transfer facility for more than 10 days must comply with the permit requirements and standards applicable to hazardous waste storage facilities (§ 263.12).

Shippers are responsible for properly preparing hazardous materials for transport. This includes proper packaging, marking, labeling, and placarding, as well as provision of the appropriate shipping paper. Generally,

- Packages must be marked with the proper shipping name of the packaged material (49 C.F.R. § 172.200).
- Most packages must also bear the consignee's or consignor's name and address (49 C.F.R. § 172.306).
- The outer packaging for liquid hazardous materials must be marked to indicate the upright position of the inside packaging (49 C.F.R. § 172.312).
- Cargo tanks and tank cars require additional markings (49 C.F.R. §§ 172.328-172.330).
- Containers and packages must have a diamond-shaped label that identifies in words and symbols the particular hazard presented by the hazardous material (49 C.F.R. §§ 172.400-.450).
- Each motor vehicle, rail car, and freight container which contains a hazardous material must be placarded on each side and end as prescribed in 49 C.F.R. §§ 172.500-.558.
- Display of identification numbers (either on the placards or on panels containing 4-in. high orange numerals) is required for portable tanks, tank cars, and cargo tanks (49 C.F.R. § 172.332).

The DOT packaging requirements are set out or referenced in the Hazardous Materials Table (49 C.F.R. § 172.101) and in Part 173 of the DOT regulations. The regulations discuss proper shipping containers in detail (49 C.F.R. § 178) and, for many substances, specify the particular containers to be used.

Shippers are required to describe on a shipping paper the hazardous material offered for shipment. DOT imposes explicit requirements for this description, including the proper shipping name, hazard class, and identification numbers listed in the Hazardous Materials table in 49 C.F.R. § 172.101. The quantity of each material must also be indicated in the description, and the shipper must certify in the shipping paper that the hazardous material has been properly prepared for transportation (49 C.F.R. § 172.202).

V. INTERIM STATUS

Any facility which treats, stores,[32] or disposes of hazardous waste (TSD facilities) must have either interim status or a permit. Facilities that do not have interim status cannot operate until they receive final permits based on the standards in Part 264. An interim status facility thus has the advantage of being able to continue treatment, storage, and disposal operations prior to issuance of a permit. During interim status, a facility must comply with the Part 265 standards which are summarized in Section VI below. In addition, the terms of a particular facility's interim status place some restrictions on changes in a facility's operations during the interim status period.

A. Eligibility for Interim Status

Under Section 3005(e) of RCRA, as amended, TSD facilities are only eligible for interim status if:

1.   They were "in existence" on November 19, 1980
2.   They have "complied with the requirements of" Section 3010(a)
3.   They have filed an application for a permit under Section 3005 (42 U.S.C. § 6925(e))

1. Definition of "Existing" Facility

An "existing" facility is one which is "in operation" or for which "construction has commenced" as of the cutoff date (§ 270.2). Construction will be considered to have commenced if the owner or operator had obtained all necessary federal, state, and local preconstruction approvals or permits; *and either*(1) a continuous physical, on-site construction program had begun, *or*(2) the owner or operator had entered into contractual obligations — which cannot be cancelled or modified without substantial loss — for physical construction of the facility to be completed within a reasonable period of time.[33] For the purposes of this provision, "Federal, State, and local approvals or permits necessary to begin physical construction" are defined as those required under federal, state, or local hazardous waste control statutes, regulations, or ordinances (§§ 260.10, 270.2).

If on or before November 19, 1980 a facility was receiving for treatment, storage, or disposal a solid waste which is subsequently listed or identified as a hazardous waste by EPA, the Agency will view the facility as having been "in existence" on that date. 45 *Fed. Reg.* 76633 (November 19, 1980). A facility that is unable to meet the "existing" facility definition cannot treat, store, or dispose of hazardous wastes.

2. Section 3010 Notification

Each person who was generating, transporting, treating, storing, or disposing of identified or listed hazardous wastes on May 19, 1980 had to file a Section 3010(a) notification within 90 days, i.e., by August 18, 1980, *regardless* of any previous notification that may have been given to the State in which the hazardous waste activity occurred. Facilities which were not in operation on May 19, 1980 were not required to file a Section 3010(a) notification by August 18, 1980.[34]

Although EPA cannot grant interim status to facilities which were required to file a § 3010(a) notification and failed to do so, the Agency has stated that it is prepared to exercise its enforcement discretion to allow such facilities to continue operating where their continued operation would be in the public interest. 45 *Fed. Reg.* 76632. EPA has issued Interim Status Compliance Letters (ISCLs) or Section 3008 compliance orders to provide formal assurances to these facilities that they will not be prosecuted for operating without a permit as long as they file a permit application and comply with the interim status standards.

3. Part A Permit Application

The third prerequisite for obtaining interim status is the timely filing of Part A of a RCRA permit application.

Part A consists of two EPA forms which were part of EPA's consolidated permit application package (§ 270.1(b)). Form 1 is the basic form which must be filled out by all applicants for RCRA permits. It requires the applicant to provide general information, including the following (§ 270.13):

*   The identification, function, and location of the facility
*   The identification, location, and ownership status of the operator

- A list of other permits
- A topographic map
- A description of activities subject to a permit requirement

EPA's Form 3 specifies additional information which is required of applicants for RCRA permits:

- The identification of the owner
- The exact location of the facility
- A decription of the processes to be used to treat, store, or dispose of wastes and the design capacities of the processes
- A listing of the hazardous wastes to be treated, stored, or disposed of and an estimate of the quantity to be treated, stored, or disposed of annually
- An indication of whether the facility is new or existing and whether this is an initial or revised application;
- For existing facilities, a scale drawing and photographs

The deadline for submission of a Part A application by facilities handling hazardous waste on May 19, 1980 was November 19, 1980. Facilities which first become subject to the hazardous waste management program as a result of a subsequent amendment to the Part 261 regulations have 6 months from the date of the amendment in which to file a permit application (§ 270.10(e)(1)).

Facilities which lose their regulatory exemption or began handling hazardous waste after November 19, 1980 can qualify for interim status if they file a permit application within 30 days of becoming subject to the hazardous waste regulations. A generator who temporarily has been accumulating hazardous waste in accordance with § 262.34, and who begins to store the waste for more than 90 days, may qualify for interim status as a storage facility if (1) the storage area was "in existence" on or before November 19, 1980; (2) the owner or operator complied with Section 3010(a); and (3) the Part A permit application is submitted within 30 days of the date the storage period exceeds the initial 90-day exemption period. 46 *Fed. Reg.* 60447 (Dec. 10, 1981).

EPA may by publication in the *Federal Register* extend the date by which a permit application must be submitted to obtain interim status for specified classes of hazardous waste management facilities upon a determination that ambiguities in the regulations have caused substantial confusion as to whether permit applications were required for such facilities (§ 270.10(e)(2)). In addition, the permit application filing deadline may be extended by a compliance order issued under § 3008 of RCRA (§ 270.10(e)(3)). Note that pending legislation, if enacted, will place additional restrictions or changes during interim status.

B. Restrictions on Changes During Interim Status

During interim status, a facility is limited to the types of hazardous wastes, the processes, and the design capacities specified in Part A of the application (§ 270.71(a)). Additional quantities of hazardous wastes may be handled without a modification to the application, provided the additional amounts do not exceed the design capacity of the facility and provided the wastes are of the same type as specified in the Part A application. New types of hazardous wastes may be handled if the owner or operator submits a revised Part A prior to the change (§ 270.72(a)). EPA approval is not required.

Increases in design capacity may be made only if a revised Part A is filed and EPA approves the change due to a lack of capacity at other hazardous waste management facilities (§ 270.72(b)). Changes in the processes employed also require a revised Part

A and EPA approval, which will be given only in an emergency situation or if necessary to comply with federal, state, or local requirements (§ 270.72(c)).

The owner or operator of a hazardous waste management facility under interim status may not reconstruct its facility without applying for a RCRA permit. The regulations provide that when the capital investment for modification exceeds 50% of the capital cost of a comparable new facility, the changes are considered a reconstruction (§ 270.72(e)).

C. Changes in Control or Ownership

A revised Part A permit application must be submitted at least 90 days in advance of any change in ownership or operational control at an interim status facility. With the exception of certain financial responsibility requirements, the obligation to comply with the interim status standards transfers to the new owner or operator as of the effective date of the change in ownership or control. The old owner or operator continues to be responsible for these financial responsibility requirements until notified by EPA. The old owner or operator is relieved of responsibility only after the new owner or operator has demonstrated compliance with these requirements (§ 270.72(d)).

D. Termination of Interim Status

Interim status exists until final administrative disposition of a permit application or until the status has been terminated. Termination may occur, after an evidentiary hearing, only for failure to furnish information required by Part B (§ 270.73). RCRA does not allow EPA to terminate interim status for violation of the substantive interim status requirements in the RCRA regulations. Instead such violations must be dealt with by enforcement actions. 45 *Fed. Reg.* 33323.

VI. INTERIM STATUS STANDARDS FOR HAZARDOUS WASTE TREATMENT, STORAGE, AND DISPOSAL FACILITIES

An interim status TSD facility must comply with the interim status standards in Part 265. The Part 265 standards contain various requirements on reporting, record-keeping, and use of the manifest system. Facilities must develop and follow certain operating procedures, such as waste analysis plans, emergency procedures, personnel training, and environmental and safety inspection schedules. Certain general standards in Part 265 apply to all interim status facilities. In addition, the regulations contain detailed standards that apply to particular types of facilities, such as landfills, storage tanks, incinerators, etc. This memorandum briefly describes the general standards and selectively highlights some aspects of the provisions for particular types of facilities.

A. Exclusions from the Part 265 Standards

Certain facilities that treat, store, or dispose of hazardous wastes are excluded from the Part 265 standards. Companies should review these exclusions carefully, as a potential means of reducing a facility's obligations under RCRA.

The Part 265 standards generally do not apply to the following hazardous waste treatment, storage, or disposal activities or facilities:

1. Ocean disposal permitted under the Marine Protection, Research, and Sanctuaries Act;
2. Underground injection permitted under the Underground Injection Control program;
3. Management of hazardous wastes by a publicly owned treatment works;
4. Hazardous waste activities subject to an authorized state RCRA program;

5. A facility permitted by a State to manage municipal or industrial solid waste, if the only hazardous waste handled by the facility falls within the small quantity generator exemption;
6. The treatment, storage, or disposal of hazardous wastes which are used, reused, recycled, or reclaimed, to the extent that such wastes are exempted by Part 261 from management as hazardous wastes;
7. Temporary storage on the site of generation in accordance with the 90-day storage exemption;
8. Disposal by a farmer of pesticides for his own use, as exempted under Part 262;
9. A "totally enclosed treatment facility";[35]
10. An "elementary neutralization unit"[36] or a "wastewater treatment unit";[37]
11. Immediate actions to contain, treat, or prevent a discharge (but Subparts C and D of Part 265 are still applicable to owners and operators of treatment, storage, or disposal facilities);
12. A transporter storing manifested shipments of hazardous waste in containers meeting the requirements of § 262.30 at a transfer facility for a period of 10 days or less; and
13. Adding absorbent material to waste in a container, or adding waste to the absorbent material in a container, provided this occurs at the time waste is first placed in the container and that the requirements for containers listed in §§ 265.17(b), 265.171, and 265.172 are met (§ 265.1(c)).

B. General Standards

All interim status TSD facilities must comply with the general requirements reviewed below.

1. Analysis of Hazardous Waste

A TSD facility must analyze a representative sample of a hazardous waste before treating, storing, or disposing of that waste in order to obtain sufficient information to handle it in accordance with Part 265 standards (§ 265.13(a)(1)). The analysis may be conducted by either the TSD facility or the generator and may use published or documented data on a particular type of waste (§ 265.13(a)(2)).

Waste analyses must be conducted in accordance with a written plan that identifies, among other things, the test methods, sampling methods, and parameters to be tested (§ 265.13(b)). Waste must be analyzed with sufficient frequency to ensure accuracy. An analysis must be repeated whenever there is reason to believe that a change has occurred in the generator's processes or materials, or that the waste does not match the description on the manifest or shipping paper. Part of the waste analysis plan must specify procedures to check upon arrival that each waste shipment matches the waste description on the accompanying manifest or shipping paper (§ 265.13(c)).

2. Security Requirements

The facility owner or operator generally must take precautions to prevent unknowing entry and minimize the possibility of unauthorized entry onto the active portions of a facility (§ 265.14(a)). A facility may use a 24-hr surveillance system, an artificial or a natural barrier, or other means to control entry to active portions of the facility at all times (§ 265.14(b)). Facilities generally must also post signs warning unauthorized personnel to keep out of active portions of the facility (§ 265.14(c)).[38]

3. Inspection Requirements

A TSD facility must develop and follow a written inspection schedule covering all monitoring equipment, safety and emergency equipment, security devices, and operating and structural equipment used to detect, prevent, and respond to hazards to human health or the environment (§ 265.15(b)(1)). The regulations also identify specific items to be inspected and minimum frequencies for these inspections. For example, areas where spills occur, such as loading and unloading areas, must be inspected daily when in use (§ 265.15(b)(4)). Remedial action must be taken to ensure that any equipment deterioration or malfunction, operator errors, or discharges do not lead to an environmental or human health hazard (§ 265.15(a), (c)). Specific information on each inspection must be recorded in an inspection log, which is to be kept for 3 years from the date of an inspection (§ 265.15(d)).

4. Personnel Training

TSD facilities must develop and implement appropriate personnel training programs to ensure that personnel are able to respond to emergencies (§ 265.16). Records must be kept describing each job, the corresponding training programs, and the training actually given to employees. Facility personnel must successfully complete a suitable training program within 6 months after the date of initial employment or transfer to a new position. Personnel must also participate in an annual review of their initial training programs.

5. Ignitable, Reactive, or Incompatible Wastes

TSD facilities must take various precautions to prevent accidental reaction or ignition of reactive or ignitable wastes. In addition, whenever specifically noted in the Part 265 standards, ignitable, reactive, or incompatible wastes must be handled so that they do not threaten human health or the environment by generating extreme heat, pressure, fire, explosion, or toxic or flammable gases or fumes (§ 265.17).

C. Preparedness and Prevention Requirements

TSD facilities must be maintained and operated "to minimize the possibility of a fire, explosion, or any unplanned sudden or non-sudden release of hazardous waste or hazardous waste constituents to air, soil, or surface water which could threaten human health or the environment" (§ 265.31). Facilities must acquire, periodically test, and maintain certain specified equipment, such as an internal communications or alarm system (§§ 265.32-.33). All personnel must have immediate access to an alarm or device capable of summoning assistance (§ 265.34).

Local police and fire departments and emergency response teams should be familiar with the TSD facility and the wastes handled at the facility. Local hospitals should be informed of the properties of the wastes handled and the injuries that could occur. TSD facilities are supposed to reach agreements establishing a primary emergency authority to cope with incidents at the facility. If a TSD facility is unable to make any of these arrangements, the refusal of state or local authorities to enter into such arrangements must be documented (§ 265.37).

D. Contingency Plan and Emergency Procedures

A TSD facility must develop a contingency plan to be implemented immediately in the event of a fire, explosion, or release of hazardous waste which could threaten human health or the environment (§ 265.51). The plan must specify the actions of all personnel in the event of such an emergency, describe arrangements with local authorities, list emergency coordinators, designate a primary emergency coordinator, list all emergency equipment and describe a personnel evacuation plan (§ 265.52).

An emergency coordinator must be either at the facility at all times or able to reach the facility in a short period of time. This coordinator must be familiar with the contingency plan and authorized to carry it out. The emergency coordinator is also responsible for complying with additional emergency procedures specified in the regulations (§§ 265.55-.56).

E. The Manifest System

TSD facilities that receive wastes from off-site must comply with the hazardous waste manifest system (§ 265.70). Upon receipt of an off-site shipment, the owner or operator (or his agent) must:

1. Sign and date each copy of the manifest to certify that the shipment was covered by the manifest;
2. Note on each copy any "significant discrepancies" between the waste shipment and the description on the manifest;
3. Give the transporter one signed copy of the manifest;
4. Send a copy of the manifest to the generator within 30 days of the date of delivery; and
5. Retain a copy of the manifest at the facility for at least 3 years from the date of delivery (§ 265.71(a)).

Slightly different requirements apply when a facility receives a bulk shipment from a rail or water transporter (§ 265.71(b)).[39]

Hazardous waste facilities must attempt to reconcile any "significant discrepancies"[40] between the waste shipment and the description on the manifest. Any significant discrepancy noted upon arrival of a waste shipment must be noted on the manifest. A discrepancy also may not become apparent until subsequent analysis of the waste. The TSD facility must attempt to reconcile the discrepancy with the generator and transporter (e.g., by phone). If a discrepancy is not resolved within 15 days, the TSD facility must immediately send a letter to EPA providing information about the discrepancy, along with a copy of the manifest (§ 265.72(b)).

A TSD facility must file a report within 15 days of receipt of any waste shipment which does not, but is required to, have a manifest. The report must identify the facility, the waste received, the method of treatment, storage, or disposal used, and the reason why the waste was not manifested (if known) (§ 265.76).[41]

F. Reporting and Record-Keeping Requirements

The owner or operator of a TSD facility must maintain an operating record at the facility that contains:

1. A description and the quantity of each hazardous waste received, and the method(s) and date(s) of its treatment, storage, or disposal
2. The location of each waste within the facility and the quantity at each location, with cross-references to the manifest document numbers; and for disposal facilities, a map of this information
3. Results of waste analyses, inspections, monitoring, and testing as required by the regulations
4. A report of incidents requiring implementation of the facility's contingency plan
5. Closure cost estimates for all TSD facilities, and post-closure cost estimates for all disposal facilities (§ 265.73)

Information must be entered as soon as it is available, and the record must be maintained until closure of the facility.

By March 1 of each even numbered year, TSD facilities must submit a detailed biennial report on waste management operations. For example, the report must identify each generator from whom waste was received, the type and quantity of waste received, the method of treatment, storage, or disposal, and the most recent closure (and if applicable, post-closure) estimates for the facility (§ 265.75).

Any required records must be available for inspection by EPA at reasonable times (§ 265.74(a)). The 3-year retention period for all records is automatically extended during the course of any unresolved enforcement action or upon request by EPA (§ 265.74(b)).

G. Compliance with Generator Standards

A TSD facility must comply with the Part 262 generator standards with respect to any hazardous wastes generated on-site (e.g., residues created by treatment processes). In addition, a TSD facility that initiates a shipment off-site of hazardous waste which was not "generated" at that facility must comply with all the generator standards with respect to that shipment (§§ 262.10(f), 265.71(c)).

H. Ground-Water Monitoring Requirements for Surface Impoundments, Landfills, or Land Treatment Facilities

Facilities which use surface impoundments, landfills, or land treatment units to manage hazardous waste are required to implement "a ground-water monitoring program capable of determining the facility's impact on the quality of ground water in the uppermost aquifer" (§ 265.90(a)). If monitoring does not reveal any ground-water quality problems, the monitoring program must be carried out during a facility's active life and during the post-closure period for disposal facilities (§ 265.90(b)). If there is a potential ground-water problem, a TSD facility may have to implement a more detailed ground-water quality assessment program in lieu of the basic ground-water monitoring program. Facilities are required to develop and maintain an outline of a suitable ground-water quality assessment program (§ 265.93(a)).

The regulations generally identify three bases for seeking a waiver of all or part of the ground-water monitoring requirements:

1. A demonstration, certified by a qualified geologist, "that there is a low potential for migration of hazardous waste or hazardous waste constituents from the facility via the uppermost aquifer to water supply wells (domestic, industrial, or agricultural) or to surface water" (§ 265.90(c))

2. A demonstration, certified by a qualified professional, that a surface impoundment, which is used only to neutralize wastes which are hazardous only due to corrosivity, will neutralize all wastes completely before they migrate out of the impoundment (§ 265.90(e))

3. A need to use an alternate ground-water monitoring system because the results of monitoring for the regularly required parameters would automatically trigger requirements for a more complex ground-water quality assessment program (§ 265.90(d))

The ground-water monitoring program requires installation of a minimum of four monitoring wells, one upgradient and three downgradient. The location, number, and depth of the wells must ensure detection of any migration of hazardous wastes or constituents to the uppermost aquifer (§§ 265.91-.92). The TSD facility must develop and follow a sampling and analysis plan to determine the concentration or value of specified parameters relating to ground-water quality.[42] The wells initially must be monitored quarterly for 1 year to establish background concentrations or values for these parameters. Thereafter, sampling and analysis must be conducted in accordance with the frequencies set out in the regulations (§ 265.92(d)).

If required comparisons of current monitoring results with background values reveal a "statistically significant difference" in the readings for downgradient wells, written notice must be given to EPA within 7 days indicating that the facility may be affecting ground-water quality (§ 265.93(d)(1)). Within the next 15 days, the facility must submit to EPA a detailed plan for a more comprehensive ground-water monitoring scheme based on the facility's ground-water quality assessment outlines (§ 265.93(d)(2)).

The ground-water quality assessment plan must be implemented as soon as possible to determine the rate and extent of migration of hazardous wastes in the ground-water. Within 15 days after making this determination, a written report must be submitted to EPA assessing ground-water quality (§ 265.93(d)(5)). If the facility owner or operator concludes that no hazardous waste from the facility has entered the ground-water, he may reinstate the basic ground-water monitoring program (§ 265.93(d)(6)). However, if hazardous waste from the facility is entering the ground-water, the more comprehensive ground-water quality assessment program must be continued until closure of the facility (§ 265.93(d)(7)).

The ground-water monitoring regulations impose a variety of additional reporting and record-keeping requiremnts on TSD facilities (§ 265.94). Facility owners and operators should also realize that ground-water monitoring data from interim status will be relied upon in the Part B permit application for a facility.

Note that pending legislation would curtail or place additional restrictions on the ability of interim status facilities to place hazardous waste in surface impoundments.

I. Closure and Post-Closure Requirements

Closure requirements apply to all TSD facilities, while post-closure requirements apply only to disposal[43] facilities (§ 265.110):

> Closure is the period after wastes are no longer accepted, during which the owners or operators complete treatment, storage and disposal operations, apply final cover to or cap landfills, and dispose of or decontaminate equipment. Post-closure is the period after closure during which owners or operators of disposal facilities must conduct certain monitoring and maintenance activities. 45 *Fed. Reg.* 33196.

Closure activities must be conducted so as to minimize the need for further maintenance and controls, and minimize or eliminate post-closure releases of hazardous wastes or constituents (§ 265.111).

1. Closure Requirements

A written plan must be kept at a TSD facility that identifies "the steps necessary to completely or partially close the facility at any point during its intended operating life" (§ 265.112). Among other things, the plan must include:

1. A description of how and when the facility will be closed and how the facility will comply with the closure requirements for particular types of treatment, storage, and disposal facilities
2. A schedule for final closure which includes the total time required to close the facility and the time required for intervening closure activities which will allow tracking of the progress of closure (§ 265.112(a))

A closure plan may be amended at any time and must be amended within 60 days of changes in operations or facility design that affect the closure plan (§ 265.112(b)).

The closure plan must be submitted to EPA 180 days before closure of a facility is expected to commence, i.e., within 30 days after the expected date of receipt of the final waste shipment. EPA will modify, approve, or disapprove the plan within 90 days, after providing the public an opportunity to submit written comments and request a hearing. A closure plan also must be submitted within 15 days after termination of interim status or after an order under Section 3008 of RCRA that the facility cease receiving wastes or close (§§ 265.112(c), (d)).

Generally, within *90 days* of receipt of the final waste shipment, all hazardous wastes in storage or treatment must be treated and disposed of on-site or removed from the site (§ 265.113(a)). All other closure activities must be completed within *6 months* of receipt of the final waste shipment, unless EPA specifically approves a longer closure period (§ 265.113(b)). EPA may delay completion of closure if there is a reasonable likelihood that another person will recommence operation of the site.

At the end of closure, all equipment and structures must be properly disposed of or decontaminated by removing all hazardous wastes and residues (§ 265.114). Certification must be given by both the facility owner or operator and an independent registered professional engineer that the facility has been closed in accordance with the closure plan (§ 265.115).

2. Post-Closure Requirements

Hazardous waste *disposal* facilities must keep at the facility a written post-closure plan which identifies the activities to be carried on after closure and the frequency of those activities. The post-closure plan must address the ground-water monitoring and reporting requirements for landfills, surface impoundments, and land treatment facilities and provide for maintenance of any monitoring and waste containment systems (§ 265.118(a)).

The post-closure care plan must be submitted to EPA at least 180 days before closure of a disposal facility is expected to commence, for example, within 30 days after the expected date of receipt of the final waste shipment (§ 265.118(c)). Within 90 days, EPA will modify, approve, or disapprove the plan, after providing an opportunity to submit written comments and request a public hearing (§ 265.118(d)). A post-closure plan may be amended at any time and must be amended within 60 days whenever changes in operations or facility design affect the post-closure plan (§ 265.118(b)).

Post-closure care must be provided for at least 30 years after closure, unless EPA grants a petition to discontinue some or all of the post-closure care requirements at an earlier date. EPA may, for cause, extend the duty to comply with these requirements beyond 30 years. In addition, the public may petition EPA to reduce or extend the post-closure care period (§ 265.118(f)).

Within 90 days of completion of closure, a facility must file with EPA and the local land authority a detailed survey plan indicating the location and dimensions of hazardous waste disposal areas (§ 265.119). The owner of property on which a disposal facility is located is required to record on the deed (or other instrument examined during a title search) a notation that will inform any potential purchaser that the land has been used to manage hazardous waste and that its subsequent use is restricted (§ 265.120). Generally, post-closure use of the property "must never be allowed to disturb the integrity" of the final cover, liner(s), or other components of a containment system, or the facility's monitoring systems.[44]

J. Financial Requirements

TSD facilities must establish two different types of financial assurance: (1) financial assurance for closure and post-closure care costs; and (2) financial responsibility for third-party liability for sudden and non-sudden accidental occurrences.

Financial assurance for closure and post-closure care costs may be established through a variety of mechanisms, including trust funds, letters of credit, surety bonds, financial tests, or insurance policies. The regulations set out detailed requirements on the use of each mechanism (§§ 265.143, 265.145). To provide the basis for this assurance, a TSD facility must have written cost estimates for closure and post-closure care. The closure cost estimate must equal the cost of closure at the point in the facility's operating life when "the extent and manner of its operation would make closure the most expensive" (§ 265.142(a)). The post-closure cost estimate is based on the annual cost for the post-closure monitoring and maintenance of the facility, times the number of years of post-closure care required (§ 265.144(a)). These estimates must be adjusted annually for inflation (§§ 265.142(b), 265.144(b)). New cost estimates must be prepared whenever a change in the closure or post-closure plans would affect closure or post-closure costs (§§ 265.142(c), 265.144(c)).

A TSD facility must have liability coverage of $1 million per occurrence, with an annual aggregate of at least $2 million, exclusive of legal defense costs, for claims arising out of injuries to persons or property due to *sudden* accidents (§ 265.147(a)). The owner or operator of a hazardous waste surface impoundment, landfill, or land treatment facility must demonstrate coverage for third-party damages caused by *non-sudden* accidental occurrences in the amount of $3 million per occurrence, with an annual aggregate of $6 million (§ 265.147(b)). The non-sudden liability coverage requirements are being phased in over 3 years (ending 1985) with the largest companies having to comply by January 1983 and the smallest by January 1985. Liability coverage for either sudden or non-sudden accidental occurrences may be demonstrated by use of liability insurance, a financial test, or a combination of these two mechanisms[45] (§§ 265.147(a), (b)).

K. Interim Status Standards for Particular Types of Facilities

The Part 265 interim status regulations contain particular standards for the following types of facilities: containers; tanks; surface impoundments; waste piles; land treatment facilities; landfills; incinerators; thermal treatment facilities; chemical, physical, and biological treatment facilities; and certain underground injection wells. Generally, these regulations establish technical requirements, additional reporting and waste analysis requirements, additional inspection routines, and additional closure and post-closure care requirements.

Certain important aspects of the technical regulations for several types of interim status facilities are noted below. Companies should check the regulations governing each type of facility.

1. Surface Impoundments

Surface impoundments must maintain at least 2 ft of freeboard in order to prevent overtopping (§ 265.222). Earthen dikes must have a protective cover, such as grass, shale, or rock, to minimize wind and water erosion and preserve structural integrity (§ 265.223). The freeboard level must be inspected once each operating day, and the dikes must be checked once a week for leakage or deterioration (§ 265.226). At closure, an impoundment must be closed as a landfill unless *all* impounded hazardous wastes, including underlying or surrounding contaminated soil, are removed (§ 265.228).

2. Waste Piles

Waste piles must be protected from wind dispersal by use of a cover or other means (§ 265.251). If leachate or run-off from the pile is a hazardous waste, the pile must: (1) be placed on an impermeable base and the leachate and run-off collected; or (2) be protected from precipitation and run-on by some other means (§ 265.253). Wastes containing free liquids may not be placed on the pile.

3. Landfills

Owners or operators of interim status landfills must divert run-on, collect run-off from active portions of the landfill, and control wind dispersal of hazardous waste (§ 265.302). Records must be kept showing the location and dimensions of each cell of the landfill, the contents of each cell, and the approximate location within a cell of each hazardous waste type (§ 265.309).

The closure and post-closure plans for an interim status landfill must specify the function and design of a final cover for the landfill and indicate how surface water infiltration and pollutant migration will be controlled and how erosion will be prevented (§ 265.310).

Bulk or noncontainerized liquid waste or waste containing free liquids may not be placed in a landfill that does not have a liner and a leachate collection and removal system (§ 265.314(a)). Alternatively, waste may be mixed with an absorbent material prior to disposal so that free liquids are no longer present. Containers holding free liquids generally may not be placed in such landfills unless they are solidified through the addition of absorbent material or are small and are overpacked in large drums filled with such materials (§§ 265.314(b), 265.316).

VII. STANDARDS FOR PERMITTED HAZARDOUS WASTE TREATMENT, STORAGE, AND DISPOSAL FACILITIES

The Part 264 standards apply to TSD facilities with Part B permits, as opposed to interim status.[46] A relatively small number of existing facilities have as yet received RCRA permits, and EPA expects the issuance of permits to take at least 5 years.

The permitting standards in Part 264 contain general facility standards and provisions relating to accident prevention, contingency planning, the manifest system, closure and post-closure, and financial requirements. These provisions (40 C.F.R Part 264, Subparts A—H) are virtually the same as the corresponding interim status standards discussed in Section VI *supra*. For an overview of these general requirements, refer to that discussion.

Part 264 also contains provisions which generally mirror the various technical regulations governing the operation of specific categories of facilities during interim status. However, the permanent standards in Part 264 impose additional technical requirements in the form of design and operating standards and performance standards for specific types of hazardous waste facilities. Because these requirements are quite detailed, this memorandum only highlights certain of the more important performance and design and operating standards. Additional guidance can be obtained from an EPA document entitled Permit Applicants' Guidance Manual for Hazardous Waste Land Treatment, Storage, and Disposal Facilities (May 1984), which is available from EPA upon request.

A. Ground-Water Protection Standards

Permitted facilities that treat, store, or dispose of hazardous waste in surface impoundments, waste piles, land treatment units, or landfills after January 26, 1983, must comply with the Part 264 ground-water protection standards. These programs are designed to protect the uppermost aquifer underlying the waste management area from contamination (§§ 264.92-.93). The Part 264 ground-water program has three phases: (1) detection monitoring; (2) compliance monitoring; and (3) corrective action.

The facility owner or operator is initially obliged to undertake a detection monitoring program to determine whether hazardous constituents are leaking into groundwater under the waste management areas (§ 264.98). Leakage is to be determined by detecting any statistically significant increase in the levels of certain indicator parameters or particular constituents specified in the permit. Monitored results from the hydraulically downgradient boundary of the waste management area are to be compared to background values for these parameters or constituents.

If leakage of hazardous constituents is detected, a compliance monitoring program must be instituted (§ 264.91(a)(1)). A compliance monitoring program evaluates whether the facility is in compliance with a ground-water protection standard, as established by EPA in the facility's permit, which sets concentration limits on particular hazardous constituents (§ 264.99(a)). The concentration limits for particular hazardous constituents are based on the background level of the constituent in the ground-water, the maximum concentration limit, if any, established for that constituent under the Safe Drinking Water Act, *or* any alternate limit which will not pose a susbtantial present or potential hazard to human health or the environment (§ 264.94).

Whenever this ground-water protection standard for the facility is exceeded, the owner or operator must undertake a corrective action program (§ 264.91(a)(2)). The purpose of such a program is to bring the facility back into compliance with the ground-water protection standard established in the permit (§ 264.100(a)). Under a corrective action program, the owner or operator must either remove the hazardous waste constituents or treat them in place (§ 264.100(b)). The elements of the corrective action program will be specified in the facility's permit. Corrective action must extend to the downgradient facility property boundary (§ 264.100(e)).

B. Design and Operating Standards for Surface Impoundments, Piles, and Landfills

The Part 264 regulations establish certain design and performance standards for surface impoundments, piles, and landfills that are used to treat, store, or dispose of hazardous wastes (Part 264 Subparts K, L, and N).

Each impoundment, pile, or landfill, except for portions on which wastes were placed prior to permit issuance, must have a liner that is designed to prevent any migration of wastes out of the facility during its active life (§§ 264.221(a), 264.251(a), 264.301(a)). For landfills and impoundments that are disposal facilities, the liner must be constructed of materials that "can prevent wastes from migrating into the liner during the active life of the facility" (§§ 264.221(a), 264.301(a)). In addition, waste piles and landfills must have leachate collection and removal systems immediately above the liner (§§ 264.251(a)(2), 264.301(a)(2)). The leachate collection and removal system must be designed and operated to function through scheduled closure of the facility (§ 264.251(a)(2)(ii), 264.301(a)(2)(ii)). In the case of a landfill, the system must be operated during the post-closure period until leachate is no longer detected (§ 264.310(b)(3)).

C. Incinerators

The Part 264 regulations also set performance standards for incinerators burning hazardous waste. Such facilities must be designed, constructed, and operated so as to achieve a destruction and removal efficiency of 99.99% for each principal organic hazardous constituent of each particular waste, as specified in its permit (§ 264.343(a)). The regulations also impose limits on emissions of hydrogen chloride and particulate matter from such incinerators (§ 264.343(b,c)). Owners and operators of incinerators must follow special procedures in obtaining a permit and must also meet certain operating and monitoring requirements (§§ 264.344, 264.345, 264.347).

VIII. RCRA PERMITS

A. Overview of RCRA Permit Program

The regulations in *Part 270* set out the basic requirements for RCRA permits: who must apply, when and how to apply, the conditions for RCRA permits, and grounds for modification or termination. *Part 124* sets forth the procedures to be followed in making decisions on permit applications under EPA-administered programs.

Since November 19, 1980, a person who treats, stores, or disposes of hazardous waste must: (1) have interim status or a RCRA permit; (2) qualify for a "permit by rule"; (3) have an emergency permit; or (4) be covered by one of the following exclusions (§§ 270.1(c)(2,3)):

1.   Generators who store hazardous waste on-site for less than 90 days under § 262.34
2.   Farmers who dispose of hazardous waste pesticides under certain conditions (see § 262.51)
3.   Persons who own or operate facilities solely for the treatment, storage, or disposal of hazardous wastes covered by the small generator exemption (see § 261.5)
4.   Owners or operators of "totally enclosed treatment facilities", "elementary neutralization units", and "wastewater treatment units" (see § 260.10)
5.   Transporters storing manifested waste shipments at a transfer facility for 10 days or less (see § 262.30)
6.   Persons adding absorbent material to waste, or waste to such material, in a container at the time the waste is first placed in the container (§ 270.1(c)(2)(vii)
7.   Persons engaged in immediate treatment or containment activities taken in response to an actual or threatened discharge of hazardous waste (see § 270.1(c)(3))

Permits by rule are established for qualifying owners or operators of publicly owned treatment works (POTWs) holding an NPDES permit, owners or operators of vessels holding an ocean dumping permit, and owners or operators of underground injection wells (§ 270.60).

EPA also has authority to issue a temporary emergency permit to treat, store, or dispose of a hazardous waste if EPA finds that such a permit is needed to cope with an imminent and substantial endangerment to human health or the environment (§ 270.61). Such a temporary emergency permit may allow treatment, storage, or disposal of hazardous waste by a nonpermitted facility or by a permitted facility whose permit does not cover the particular activity or waste. The permit may be oral or written and must not exceed 90 days.

The operator of a hazardous waste facility is responsible for applying for and obtaining a permit (§ 270.10(b)). However, when the facility is owned by a person other than the operator, the owner must also sign the permit application. (*Id.*)[47] In addition, it is EPA's position that such an owner is jointly responsible with the operator for compliance with the permit. 45 *Fed Reg.* 33295 (May 19, 1980).

1. Signatories

Permit applications submitted by corporations must be signed by a principal executive officer of at least the level of vice president (§ 270.11(a)(1)).[48] The signatory must also certify that he has personally examined and is familiar with the information submitted (§ 270.11(d)).

2. Contents of Part B

Companies should not underestimate the time, advance planning, and costs involved in completing a Part B permit application. The application must be in narrative form, supplemented with the requisite maps, diagrams, plans, and facility designs. The regulations do not specify a particular form or sequence for the Part B application, but EPA recommends that applicants follow the Agency's guidance documents and permit application checklist. Use of outside consultants or counsel may relieve companies of some of the burden of identifying currently applicable regulations and compiling and organizing the required information.

The requirements for Part B call for information concerning virtually *all* aspects of a facility (§ 270.14) and generally require specific information regarding compliance with the Part 264 substantive standards. Written plans required under Part 264 generally must be submitted as part of a Part B application. Companies are cautioned that certain elements required for a Part B application may not be readily available, e.g., a wind rose (see § 270.14(b)(19)(v)).

Facilities subject to the Part 264 ground-water monitoring program must provide detailed information about the location of monitoring wells, the proposed point of compliance, ground-water flow and direction, any existing ground-water contamination, and potential indicator parameters (§ 270.14(c)). Additional information is requested for permits for particular types of facilities: containers (§ 270.15); tanks (§ 270.16); surface impoundments (§ 270.17); waste piles (§ 270.18); incinerators (§ 270.19); land treatment facilities (§ 270.20); and landfills (§ 270.21).

3. Time for Filing Permit Applications

For existing facilities, Part A of the RCRA application must be filed, or have been filed, within 6 months of the date of publication of regulations which first required the applicant to comply with the interim status standards (§ 270.10(e)(1)). Part B of the application is generally due 180 days after EPA notifies individual existing facilities that they must submit Part B. Existing facilities may also submit Part B voluntarily (§ 270.10(e)(4)).

Permit applications for new hazardous waste facilities must be submitted at least 180 days before construction is expected to commence. Physical construction on a new facility cannot actually begin until a finally effective RCRA permit has been issued (§ 270.10(f)).

4. Records Relating to a Permit Application

A permit applicant is required to keep records of "all data" used to complete permit applications and any "supplemental information" submitted under the permit application provisions for 3 years from the date of signing the application (§§ 270.10(i), 270.30(j)(2)). During the term of the permit, a permittee is required to inform EPA at any time of a discovery that "it failed to submit any relevant facts in a permit application, or submitted incorrect information in a permit application or in any report" (§ 270.30(l)(11)).

B. RCRA Permit Terms and Conditions

RCRA permits will be issued for a fixed term which may not exceed 10 years,[49] although the permit issuer may set the term at less than 10 years (§ 270.50).

1. Permit Modification

Normally, if a facility is in compliance with the terms of its permit, the permit will not be modified. However, a permit can be reopened for cause for a "major modification". A major modification can be required due to:

1. Alterations or additions to a facility that justify additional or changed permit conditions
2. New information received by EPA which was not available at the time of permit issuance but which would have justified the application of different conditions
3. A timely request by a permittee to modify the permit in light of regulatory or judicial action subsequent to issuance

4.   A need to modify a compliance schedule due to events over which permittee has little or no control
5.   A need to alter certain terms of the permit, to reflect various Part 264 requirements, as identified in the regulations (§ 270.41(a))

A permit may be modified when cause for termination exists, but EPA decides to modify, rather than terminate, the permit. Transfer of a permit may also result in modification (§ 270.41(b)).

If a permit is reopened for a "major modification", then the permit must be modified in accordance with certain procedural requirements in Part 124, discussed *infra*, which call for a draft permit and public review (§ 270.41). The regulations also identify certain "minor modifications" that do not trigger the Part 124 procedural requirements. Among other things, a minor modification to a permit can:

1.   Make minor changes in interim dates in a compliance schedule
2.   Change certain conditions for trial burn and incineration permits
3.   Require more frequent monitoring or reporting
4.   Transfer ownership or operational control based on an appropriate written agreement
5.   Change certain conditions for land treatment units and
6.   Change estimates of maximum inventory or expected closure schedules (§ 270.42)

A permit that is subject to a modification proceeding is reopened only as to the conditions to be modified (§§ 124.5(c)(2), 270.41).

2. Revocation and Reissuance

EPA has discretion to revoke and reissue a permit (in lieu of modification) in two circumstances: (1) if cause for termination exists; and (2) upon receipt of a notification of a proposed transfer of a permit (§ 270.41(b)). In addition, where cause exists for modification, a permittee may request that the permit be revoked and reissued (§ 270.41(a)). When a permit is revoked and reissued, the *entire* permit is reopened and is subject to revision. The permit is reissued for a new term.

3. Termination

Permits may be terminated only for cause and only after notice and an opportunity for a hearing. However, the grounds for termination are extremely broad. For example, a permit may be terminated upon a finding that "the permitted activity endangers human health or the environment and can only be regulated to acceptable levels by permit modification or termination" (§ 270.43(a)(3)). Other causes for termination include noncompliance with any condition of the permit and misrepresentation or omission of any "relevant facts" during the permit-issuance process (§ 270.43(a)(1), (2)).

4. Transfer

RCRA permits may generally be transferred to a new owner or operator only by modification or revocation and reissuance of the permit (§ 270.40). Thus, new owners or operators should not assume that the terms and conditions of the prior permit will remain unchanged since the entire permit is subject to review upon revocation.

5. General Conditions

A number of general conditions apply to all RCRA permittees, such as the duty to comply with permit conditions, to reapply 180 days prior to expiration, to maintain and operate facilities properly, and to mitigate any adverse impact resulting from non-compliance (§ 270.30). The permittee has duties to supply information to the permit-issuing authority to allow inspection of the facility, to provide access to records, and to allow sampling or monitoring (§§ 270.30(h), (i)). Records on monitoring activities must be kept for 3 years (§ 270.30(j)(2)). Notice must be given, or reports made, regarding any planned physical changes to the permitted facility, any anticipated noncompliance, monitoring results, compliance schedule progress, and instances of noncompliance (§ 270.30(k)(1)).

In addition, noncompliance which may endanger health or the environment must be reported orally within 24 hr and in writing within 5 days of the time the permittee becomes aware of it. A RCRA permittee must report orally within 24 hr of the receipt of information concerning releases, discharges, or accidents which may endanger public drinking water, the environment, or human health outside the facility (§ 270.30(l)(6)).

6. Compliance Schedules

RCRA permits may contain schedules for achieving compliance (§ 270.33). If the period for compliance exceeds 1 year, the schedule must contain interim requirements and dates (§ 270.33(a)(2)). Within 14 days following each date, the permittee must file a report of compliance or noncompliance with the schedule. If the permittee decides to terminate operations, the permit may include a final schedule of compliance. If the permittee is undecided whether to terminate or to continue, the permit may include two alternative schedules — one leading to compliance and one leading to termination (§ 270.33(b)).

C. Procedures for Permit Decision-Making

40 C.F.R. Part 124 sets out procedures to be followed in the processing of applications and in the modification, revocation and reissuance and termination of permits.

1. Initiation of Permitting Process

Upon receipt of a RCRA permit application, EPA must initially review the application for completeness and notify the applicant of any deficiencies.[50] Failure or refusal to correct deficiencies is a basis for permit denial. This review is to be completed within 30 days for a new facility and 60 days for an existing facility (§ 124.3(c)), but EPA is having difficulty meeting these deadlines. Once the completeness review is over, EPA is not subject to any time limits in making a substantive evaluation of the application.

Requests for modification, revocation and reissuance, and termination are subject to essentially the same permit procedures as permit applications (§ 124.5).[51] EPA must prepare draft permits for a tentative decision to issue, modify, or revoke and reissue a permit (§§ 124.5(c), 124.6(d)). A notice of intent to deny must be prepared if EPA tentatively decides to deny a permit application (§ 124.6(b)). A notice of intent to deny and a draft permit are subject to the same review and comment procedures. *Id.*

Draft permits must include all conditions, requirements, and technical standards to be included in the final permit and must be accompanied by either a statement describing the basis for the draft or a fact sheet[52] (§§ 124.6(d,e)). The administrative record, which EPA will use to support the draft permit, consists of: (1) the application and supporting data; (2) the draft permit itself; (3) the statement of basis or fact sheet and all documents cited therein; and (4) other documents in the supporting file (§ 124.9).

Public notice must be given, pursuant to specific regulatory requirements (§ 124.10), of a draft permit or a notice of intent to deny. The notice must allow at least 30 days for interested parties to submit comments, and any notice announcing a public hearing must be issued at least 30 days before the hearing (§ 124.10(b)). If no public hearing is announced, interested members of the public may request a public hearing during the comment period (§ 124.11).

A public hearing must be held whenever EPA receives written notice of opposition to a draft permit and a request for a hearing within 45 days of public notice. EPA may hold public hearings on proposed permit actions (other than termination) if the Agency decides that there is a significant degree of public interest in a draft permit or that a hearing would clarify the issues or otherwise be beneficial.[53] The regulations prescribe informal hearing procedures for the submission of oral or written statements and data by any interested parties (§ 124.12).

Anyone who disagrees with the conditions of a draft permit, or with a tentative decision to deny or to terminate a permit or interim status "must raise all reasonably ascertainable issues and submit all reasonably available arguments and factual grounds supporting their position, including all supporting material" during the public comment period or at the public hearing if one is held (§ 124.13).[54] Thus, a permit applicant may only have the 30-day comment period in which to identify and document his objections to the draft permit, unless a hearing is held.

Companies may be able to minimize the burden of responding to a draft permit by soliciting EPA's views through informal meetings prior to submission of the permit application. In addition, commenters may request a longer comment period (§ 124.13). If substantial new questions are raised during a comment period, EPA may revise the draft permit, prepare a revised fact sheet or basis, or reopen the comment period (§ 124.14).

After the comment period (and any public hearing), EPA will issue a final decision and will notify all who commented or requested notice of the final decision (§ 124.15). EPA will also issue a response to comments (§ 124.17) and compile the administrative record of the final permit decision.

A final permit decision generally is effective 30 days after notice of the decision, unless: (1) a later effective date is specified; (2) an appeal is filed; or (3) an evidentiary hearing is requested (for permit termination only). If no comments were filed requesting a change in the draft permit, then a permit is immediately effective upon issuance (§ 124.15(b)).

Within 30 days after service of a final permit decision, any person who filed comments or participated in a hearing may appeal the permit terms by petitioning the Administrator for review of the permit decision (§ 124.19). Other persons may petition only for review of differences between the final decision and the draft. The petition must state the reasons for review, including a showing that the terms to be reviewed are based on clearly erroneous conclusions of law or fact or involve an important policy issue warranting review (§ 124.19(a)). EPA may also act on its own initiative during this 30-day period to review any permit condition (§ 124.19(b)). Appeal of permit terms by filing a petition with the Administrator is a prerequisite to judicial review (§ 124.19(e)).

When an appeal to the Administrator has been taken or an evidentiary hearing granted in a consolidated RCRA-NPDES proceeding[55] the contested permit conditions are stayed pending final action. Uncontested conditions which are not severable from contested conditions also are stayed. EPA will notify all parties of any stayed permit conditions. All other conditions shall remain fully effective and enforceable (§ 124.16(a)).

In the case of a renewal of an existing permit, the uncontested terms of the new permit and all other terms and conditions of the existing permit remain in effect during the evidentiary hearing or the appeal to the Administrator. The regulations do not address whether an applicant for an initial RCRA permit may continue to operate under interim status during an administrative appeal of or evidentiary hearing on the permit. However, it would seem reasonable that, at least regarding contested terms, the conditions of interim status would continue to apply.

2. Evidentiary Hearing Procedures

An opportunity for an evidentiary hearing is available in the case of termination or suspension of a RCRA permit or termination of interim status for failure to file an adequate permit application (§ 124.71(a)). Issuance, modification, or revocation and reissuance of RCRA permits are not subject to an evidentiary hearing unless the conditions of such permits are linked with the conditions of an NPDES permit to which an evidentiary hearing has been granted and the proceedings have been consolidated (§ 124.71(b)).

a. Request for Evidentiary Hearing

A request for an evidentiary hearing on a RCRA permit or interim status termination must be submitted within 30 days after service of EPA's decision to terminate (§ 124.74(a)). For a RCRA permit to be included in an evidentiary hearing on an NPDES permit, a request must be filed with the EPA Regional Office within 30 days after EPA has served notice of issuance of the NPDES final permit. The Regional Administrator may extend the time allowed for good cause (§ 124.74(f)).

The regulations are ambiguous as to the scope of a hearing on a RCRA or UIC permit which has been consolidated with an NPDES permit. The regulations state that persons "requesting an evidentiary hearing on an NPDES permit under this section may also request an evidentiary hearing on a RCRA or UIC permit" if:

1. The processing of the RCRA and NPDES permits have been consolidated;
2. Modification of the RCRA permit is likely to be needed as a result of resolution of the NPDES permit issues; and
3. The standards for granting a hearing on the NPDES permit are met (§ 124.72(b)(2)).

There is no provision in the regulations stating that only those conditions of the RCRA permit which are interrelated with the NPDES permit may be contested in an evidentiary hearing. It would seem advisable, nonetheless, to point out such interrelations to the extent that they exist.

Since an evidentiary hearing request is a necessary step to preserve the right to judicial review, a company in consolidated NPDES and RCRA proceedings should file an evidentiary hearing request on all conditions of the RCRA permit considered objectionable, file the necessary supporting information for that portion of the request as well as for the request on the NPDES permit, and demonstrate to the extent possible the interrelations between the NPDES permit and the contested provisions of the RCRA permit. In addition, it may be prudent to file an appeal with the Administrator as a safeguard in the event that EPA decides that the RCRA permit is not so closely related to the NPDES permit as to require an evidentiary hearing on the RCRA permit. There appears to be no such ambiguity with respect to termination of a RCRA permit or interim status. All aspects of such termination are proper subjects of an evidentiary hearing.

The regulations set forth the elements of a request for an evidentiary hearing. The request must include, among other things, the following:

1. A statement of the legal and factual issues and their relevance to the permit decision
2. Designation of the specific factual issues to be adjudicated

3.   Information or written documents supporting the request
4.   An estimate of the hearing time required
5.   A statement of the nature and the scope of the requester's interest
6.   An agreement to be available as a witness and to make available as witnesses all persons represented by the requester and "all" officers, directors, employees, consultants, and agents of both the requester and persons represented by him
7.   Specific reference to the conditions of the permit contested and suggested alternative conditions
8.   If the technology upon which the permit is based is being challenged, identification of the basis for the challenge and suggested alternative technologies
9.   Identification of obligations under the permit which should be stayed if the request for a hearing is granted (§ 124.74(b), (c))

Documents must be submitted to support the request. However, the support should be sufficient to convince EPA that a material issue of fact exists and to provide a basis, together with the administrative record, for an appeal to the Administrator or to the courts if the request for a hearing is denied.

While not explicitly required by the regulations, a request for an evidentiary hearing should include grounds constituting "good cause" for raising issues or relying on evidence not submitted during the public comment and hearing period (§ 124.76). Good cause exists when the party seeking to submit evidence or raise issues could not reasonably have ascertained the issues or made evidence available or could not have reasonably anticipated their relevance or materiality. In the absence of a demonstration of good cause, the Regional Administrator is likely to deny the request for an evidentiary hearing on issues or evidence not raised earlier.

A key to preparation of an adequate request for an evidentiary hearing will be the administrative record compiled during the public comment and hearing period. At a minimum, the permit applicant should obtain EPA's response to comments submitted during the public comment and hearing. In addition, it may be necessary in many instances to review the entire administrative record compiled by EPA in connection with the issuance of the permit.

b. Decision on the Request

The Regional Administrator must issue a decision on the request for an evidentiary hearing within 30 days following the expiration of the time allowed for submission of the request (§ 124.75(a)(1)). The regulations do not provide specific criteria for the grant or denial of a request, but provide that it may be denied if the request is incomplete or fails to set forth material issues of fact relevant to the issuance of a permit. *Id.* If an evidentiary hearing is denied, an appeal to the EPA Administrator is available and is a prerequisite to judicial review (§§ 124.60(g), 124.75(b), 124.91(f)). If an evidentiary hearing is partially granted and partially denied, the regulations apparently require an immediate appeal to the Administrator of the portions denied (§ 124.75(b)).

EPA will give public notice of the grant of an evidentiary hearing. At the same time, it will designate the Agency trial staff which will be responsible for investigating, litigating, and presenting evidence on behalf of EPA (§§ 124.77, 124.78(a)(1)).

Any person may be admitted as a party to the evidentiary hearing on the basis of a request filed within 15 days after issuance of public notice if the request meets the requirements for an evidentiary hearing request and if the person identifies the issues which he wishes to address (§ 124.79(a)). Thereafter, third parties may intervene only in extraordinary circumstances, and an intervenor may raise new issues only for "good cause" (§ 124.79(b)).

c. Legal Issues

EPA's commentary indicates that the Regional Administrator will deny an evidentiary hearing if only purely legal issues are raised, thus allowing the requester to appeal such issues to the Administrator (§ 124.74, (Note)). It appears possible that a permit applicant may have to pursue an appeal to the Administrator on certain legal issues while an evidentiary hearing is proceeding on other issues.

d. Addition of Issues and Conditions

The regulations provide that only contested terms of a permit will be "affected" by or "considered" at an evidentiary hearing (§ 124.74(d)). This provision generally should preclude EPA from seeking to add new provisions or modifying uncontested provisions during the evidentiary hearing process.

e. Prehearing Procedures

Once the parties to and issues in an evidentiary hearing have been identified, the Regional Administrator will request that the Chief Administrative Law Judge appoint an administrative law judge as Presiding Officer (§ 124.81). The Presiding Officer may take a wide range of actions with respect to scheduling, ruling on the admissibility of evidence, simplifying or clarifying issues, or expediting the hearing (§ 124.85).

The regulations provide for summary determinations as a principal means of resolving issues. Any party may file a motion for summary determination, generally at least 45 days before the date set for the hearing, on the ground that there is no genuine issue of material fact to be resolved. Responses or countermotions must be filed within 30 days after service of the motion for summary determination. If the motion was supported by affidavits or other factual materials, the response must include such support (§ 124.84).

The Presiding Officer has 30 days after the date on which responses are filed to grant or deny the motion for summary determination (§ 124.84(d)). At the request of a party, the Presiding Officer may certify the question for interlocutory appeal to the Administrator (§ 124.90(a)). The Presiding Officer's refusal to certify can be appealed to the Administrator only when an appeal is shown to be "in the public interest" (§ 124.90(c)).

f. Burden of Going Forward and Burden of Persuasion

The regulations provide that the applicant always has the burden of persuading EPA to issue a permit (§ 124.85(a)(1)). EPA has the burden of going forward to present an "affirmative case in support" of any contested term of the permit it wrote (§ 124.85(a)(2)). At the same time, any party who argues that particular terms should be deleted or that other terms should be added has the burden of going forward with an affirmative case on those questions (§ 124.85(a)(3)). Because EPA expects that its burden of presenting an affirmative case will be met by its administrative record, including the fact sheet and response to comments (§ 124.85(a)(1), Note), it is reasonably clear that EPA expects the applicant to bear both the burden of initially submitting evidence on the issues raised in the request for an evidentiary hearing and the ultimate burden of proof.

g. Conduct of Evidentiary Hearing

The Presiding Officer is granted considerable discretion in the conduct of an evidentiary hearing consistent with the possibly conflicting goals of giving parties an opportunity for a full and impartial hearing and reaching an expedited decision (§ 124.85(b)). In particular, the Presiding Officer has authority to regulate cross-examination. Pursuant to the regulations, EPA witnesses may not be cross-examined on "questions of law" or "policy", or on matters not subject to challenge in an evidentiary hearing. The party seeking cross-examination has the burden of demonstrating that cross-examination is likely to result in clarifying or resolving an issue and that the issue cannot be better clarified in other ways (§ 124.85(b)(16)).

EPA has imposed a greater burden of production upon parties requesting an evidentiary hearing than it has on itself. However, the provision of the regulations authorizing the Presiding Officer to order that "sponsoring witnesses" be provided for cross-examination on the portions of the administrative record for which they were responsible (§ 124.85(d)(2)) may provide a method by which EPA can be compelled to provide witnesses to support the administrative record (see § 124.85(b)(16)).

The regulations provide that all direct and rebuttal evidence must be submitted in written narrative form in the absence of good cause (§ 124.85(c)). The Presiding Officer must admit all relevant, competent and material evidence, unless it is repetitious (§ 124.85(d)).

Rulings may be certified by the Presiding Officer for interlocutory appeal to the Administrator (§§ 124.85(d)(5), 124.90). In the absence of such certification, adverse rulings of the Presiding Officer may be an issue during administrative and judicial review following the hearing (§ 124.90(e)).

h. Initial Decision and Appeal

After receipt of proposed findings of fact, conclusions of law, and briefs from the parties (§ 124.88), the Presiding Officer will prepare an initial decision on the issues raised in the hearing (§ 124.89). The Regional Administrator does not review that initial decision. Rather, any party, including EPA, may file an appeal with the Administrator within 30 days after service of the Presiding Officer's decision. The appeal petition should include, wherever appropriate, a showing that the Presiding Officer's findings of facts are clearly erroneous or that an important issue of policy is involved which should be decided by the Administrator (§ 124.91(a)(1)). In granting a petition to review, the Administrator may limit the issues to be reviewed (§ 124.91(c)(1), (d)). The Administrator, presumably at the request of the Office of Enforcement, may review an initial decision on his own initiative (§ 124.91(b)). The Administrator may decide the appeal himself or appoint a Judicial Officer (§ 124.72(b)).

The procedures and deadlines for briefs and other appeal filings are set forth in the regulations (§ 124.91(g)). An appeal to the Administrator is a prerequisite to judicial review (§ 124.91(e)).

i. Other Matters

The regulations contain a number of additional provisions of which the applicant must be aware. These include requirements as to format and content of documents submitted for the record (§ 124.73), requirements for filing and service of written submissions (§ 124.80), protection of confidential information submitted for the record (§ 124.85(b)(15)), and restrictions on ex parte communications (§ 124.78).

3. Panel Procedures

The regulations provide three instances in which the "nonadversary" panel procedures can be used for processing draft RCRA permits. First, the Regional Administrator may decide as a matter of discretion to hold a panel hearing. Second, a party entitled to an evidentiary hearing may request a panel hearing instead. Third, the Regional Administrator may elect to apply the panel hearing procedures to an initial licensing of an NPDES facility, in which case a consolidated RCRA draft permit would also be processed under those procedures (§ 124.111).

a. Public Hearing and Comment Period

The process is initiated with the formulation of a draft permit and a public notice with a 30-day comment period and, at the discretion of the Regional Administrator, a public hearing (§ 124.113). However, the Regional Administrator may decide that a panel hearing will be necessary and include that determination in the public notice of the draft permit (§ 124.114(c)). Both a public and a panel hearing may be held.

b. Request for a Panel Hearing

If the initial public notice does not call for a panel hearing, the permit applicant or other person must file a request for a panel hearing prior to the close of the public comment period (§ 124.114(a)). The required content of a request for a panel hearing is similar to that for an evidentiary hearing. *Id.* If the Regional Administrator decides at the outset to hold a panel hearing, it appears that the permit applicant would have to file a request to participate in the hearing by the deadline specified in EPA's notice (§ 124.117).

c. Submission of Comments

The applicant and other participants in a panel hearing must submit their written comments and supporting evidence prior to the hearing. EPA expects that written comments will constitute the bulk of the record and will try to limit oral testimony and the hearing to arguments emphasizing the major points in the comments or to points that could not have been made in the written comments (§ 124.118). In most cases, parties to a panel hearing will be restricted in the evidence which may be presented to that submitted during the initial public comment period, as is the rule for evidentiary hearings (§§ 124.112(e), 124.13, 124.76).

d. Hearing Procedures

An administrative law judge or, if the parties agree, an EPA lawyer without prior connection to the matter, will serve as the Presiding Officer. The Presiding Officer's authority to conduct a panel hearing is substantially the same as the Presiding Officer's authority in an evidentiary hearing (§ 124.119).

The panel for a hearing will consist of no fewer than three EPA employees having "special expertise in areas related to the hearing issue," only one of whom may have had a hand in preparing the draft permit (§ 124.120(a)). EPA may change the panel membership during the course of the proceeding and add persons (such as consultants) who are not EPA employees. At the discretion of the Regional Administrator, EPA may appoint a trial staff to represent the Agency's point of view (§ 124.120(b)).

The panel members may cross-examine witnesses during the first, or legislative, phase of the hearing (§ 124.120(d)). Other participants may cross-examine at this phase only if the Presiding Officer finds that it would expedite consideration. *Id.* Participants may also submit written questions (§ 124.120(e)).

If the Presiding Officer grants a written request filed after the transcript of the first phase of the hearing is filed, parties may also conduct oral cross-examination (§ 124.121). If granted, a second or adjudicatory phase of the hearing will be held. A participant also may request that sponsoring witnesses be provided for portions of the administrative record (§ 124.121(f)).

The burden of persuasion and the burden of presenting an affirmative case are the same in a panel hearing as they are in an evidentiary hearing (§ 124.112(d)(5)).

4. Decisions and Appeal

Following the hearing, the participant may file proposed findings of fact, conclusions of law, and supporting briefs (§ 124.123). A person will be named to prepare a recommended decision in consultation with panel members (§ 124.124).

The parties have 30 days after service of the recommended decision to file an appeal with the Administrator (§ 124.125). The procedures for appeal from a panel hearing are identical to those for an appeal following an evidentiary hearing. Further, the procedures for interlocutory appeal are applicable to panel hearings (§§ 124.112(a)(8), 124.90).

A final decision will be issued by the Administrator ''[a]s soon as practicable'' in the event of an appeal (§ 124.126). If there is no appeal, the recommended decision becomes final 30 days after service (§ 124.127, § 124.125).

IX. STATE HAZARDOUS WASTE PROGRAMS UNDER RCRA

In enacting RCRA, Congress intended to encourage states to develop and implement hazardous waste programs equivalent to the federal program. Congress also required that existing state programs be phased in by providing for interim authorization for state programs which were in existence less than 90 days after the promulgation of regulations under RCRA. Beginning 180 days after the promulgation of regulations under RCRA, those programs are to be granted interim authorization so long as they are ''substantially equivalent'' to the federal program (RCRA § 3006(c)).

To obtain Phase I interim authorization, the state program must include the following (§ 271.128):

1. Provisions for identification and listing of hazardous wastes — These provisions must call for the control of wastes ''nearly identical'' to those controlled by EPA.
2. Manifest system and related generator and transport requirements — If a state does not have an equivalent system, these portions of the federal program will be operated in the state, and interim authorization may still be issued for other parts of the program.
3. Requirements of hazardous waste treatment, storage, and disposal facilities — The state regulations must be ''substantially equivalent'' to Part 265 of EPA's regulations. States may include hazardous waste injection wells, which would qualify for interim status under the EPA program, within the interim authorization program until a state UIC program has been approved.
4. Enforcement authority — The states' enforcement authority must be comparable
5. Compliance evaluation program — These provisions must be comparable to corresponding provisions in EPA's program.

For Phase II of interim authorization, the state program must, in addition, have the following (§ 271.129): (1) standards for hazardous waste management facilities providing protection substantially equivalent to Part 264 of EPA's regulations, and (2) a permit program.

EPA has promulgated regulations designed to assure that state programs provide what it deems a minimum level of protection from hazardous wastes, while encouraging as many states as possible to assume responsibility for RCRA programs (40 C.F.R. Part 271). These regulations contain the mechanism whereby states can obtain (1) interim authorization and (2) final authorization to operate hazardous waste programs.

A. Interim Authorization of State Hazardous Waste Programs

Under RCRA, state programs may be instituted under a preliminary transitional authorization prior to final authorization (Subpart B; § 271.121). The regulations provide that a state may obtain interim authorization in two phases — Phase I and Phase II, which correspond to the phases in the federal program — extending for no more than 24 months beyond the effective date of the last component of Phase II (§

271.122(b)(1)). At the expiration of the 24-month period, if no final authorization has been granted, EPA will administer the federal program in the state (§ 127.122(b)(2)). Phase II authorization is further divided into Components A, B, and C which correspond to one or more specific categories of facilities which require RCRA permits.[56] See 46 *Fed. Reg.* 8298 (Jan. 26, 1981).

States desiring interim authorization must submit an application which demonstrates that the state agency has the requisite authority, that the program complies with EPA requirements, that EPA's oversight responsibility is recognized, and that the state will proceed to develop a program meeting the requirements for final authorization (§§ 271.123—271.127).

Within 30 days of receipt of a state program submission, EPA must issue a public notice regarding the program and allow 30 days for comment. Within 90 days of the public notice, EPA must make a decision on the program (§ 271.135). The regulations also provide that if it appears that a state will not develop its own program for final authorization, EPA may withdraw interim authorization and proceed to regulate activities in the state (§ 271.136).

If the state fails to file an application for Phase II interim authorization or if its application fails to meet the requirements, then the state program will be terminated and the administration of RCRA in the state will revert to EPA (§ 271.137).

EPA retains the right to bring an enforcement action for noncompliance with (1) the conditions of any permit issued under state interim authorization or (2) any condition which EPA stated (in comments on the draft permit) was necessary to implement program requirements regardless of whether the condition was included in the state-issued permit (§ 271.134(e)).

Most states have obtained Phase I interim authorization. A lesser number of states have obtained various components of Phase II authorization.

B. Final Authorization of State Hazardous Waste Programs

To be approved for final authorization, state programs must be consistent with the federal program and must contain certain minimum requirements. To the extent that a state program unreasonably impedes the interstate movement of hazardous wastes for authorized treatment, storage, or disposal or prohibits treatment, storage, or disposal in the state without basis in human health or environmental protection, it shall be deemed inconsistent and will not be approved (§ 271.4(a) and (b)). Any state program that does not contain the manifest system requirements is also inconsistent. State programs that are overly stringent in these areas thus may be rejected by EPA under the "consistency" requirement.

On the other hand, the regulations provide that, except for these consistency provisions, nothing precludes a state from adopting or enforcing more stringent requirements than those set forth in the federal regulations or operating a program with a broader scope of coverage than that required under the federal rules (§ 271.1(i)).

Certain specific requirements must be included in state permit programs in order to obtain final authorization. Among the requirements are the following (§ 271.14):

1. Types of facilities specifically included in the program (§ 270.1(c)(1))
2. Effect of permit (§ 270.4, 270.1(c))
3. Application procedures and informational requirements (§§ 124.3(a), 270.10-.29)
4. Monitoring and reporting requirements (§§ 270.5, 270.31)
5. Permit conditions (§§ 270.30, 270.32)
6. Schedules of compliance (§ 270.33)
7. Permit transfer, modification and termination (§§ 270.40, 270.41, 270.43)
8. Duration (§ 270.50)
9. Permits by rule and emergency permits (§§ 270.60-.61)

10.   Basic procedures for permit decision-making (§§ 124.5, 124.6, 124.8, 124.10, 124.11, 124.12, 124.17 (in part)).

The states have considerable latitude as to the procedures they adopt for the issuance of permits. They must follow the requirements for public notice, public comment, public hearing, preparation of fact sheets and statements of basis for draft permits, and preparation of a response to public comments. Beyond those minimum requirements, any procedural rights accorded to a permit applicant are a matter of state law.

EPA will require that, before final authorization, states have enforcement capability similar to EPA's. A state must have authority to seek civil penalties, injunction, and certain criminal sanctions (§ 271.16).

The regulations provide that EPA may comment on permit applications and draft permits following final authorization (§ 271.19(a)). Enforcement actions may be brought by EPA for noncompliance with actual conditions of a permit (§ 271.19(e)). If the Regional Administrator states that certain conditions in a permit are "necessary" to implement the state program, EPA may bring an enforcement action for noncompliance with such conditions, even if they were never included in the permit (§ 271.19(e)(2)). In addition, EPA can bring an imminent hazard action pursuant to Section 7003 of RCRA (§ 271.19(e)(4)).

EPA also may inspect generators, transporters, and HWM facilities, but may agree with the states to limit such inspections for all facilities except major HWM facilities (§ 271.8(b)(4)).

Only a small number of states have applied for final federal authorization of their hazardous waste permit programs under RCRA.

FOOTNOTES

[1]   The regulations discussed in this memorandum reflect the basic federal hazardous waste program. Under RCRA, states are allowed to develop and enforce requirements more stringent than the federal regulations.

[2]   Except as otherwise noted, regulations cited in this memorandum refer to provisions in Title 40 of the Code of Federal Regulations.

[3]   EPA has proposed substantial amendments to the definition of "solid waste" which replace the "sometimes discarded" test with "discarded material". The Agency hopes to issue final regulations in 1984. (See 48 *Fed. Reg.* 14472 (April 4, 1983).

[4]   Under the regulations, the fact that these materials are sometimes recycled or reclaimed does not keep them from being considered either solid or hazardous wastes.

[5]   Industrial waste-water which is being stored or treated prior to discharge to navigable waters or to a public sewer treatment system and sludge generated by industrial waste-water treatment are still within the definition of solid waste. See comment to § 261.4(a) (2).

[6]   *Id.*

[7]   These wastes have been listed because they either meet one of the four characteristics in Subpart C or meet the criteria for toxicity. Appendix VII to Part 261 identifies the particular constituent which caused wastes to be listed on the basis of toxicity. The wastes on these two lists are currently subject to the 1000 kg/month exemption level for small quantity generators. Note, however, that legislation is pending in Congress to narrow substantially the small quantity generator exemption. See Section II.F.2 *infra*.

[8]   A listing of a chemical under its generic name includes all products in which it is the sole active ingredient. Since EPA considers materials on the third list to be acutely hazardous, the small generator exemption level for materials on this list has generally been set at 1 kg/month. Wastes on the fourth list are generally subject to the 1000 kg/month exemption level for small quantity generators currently in effect.

[9]   See comment to § 261.33(d).

[10] Such a process waste, however, may still be a hazardous waste if it meets any of the characteristics of hazardous waste discussed in Section II.E *infra*.

[11] EPA has defined discharge as "the accidental spilling, leaking, pumping, pouring, emitting or dumping of hazardous wastes or materials which, when spilled, become hazardous wastes into or on any land or water" (§ 260.10).

[12] "Representative sample" is defined as a "sample of a universe or whole (e.g., waste pile, lagoon, ground water) which can be expected to exhibit the average properties of the universe or whole" (§ 260.10). A representative sample may be obtained by any of the methods in Appendix I to Part 26 (§ 261.20(c)), or by other method which meets this definition.

[13] Where the waste contains less than 0.5% filterable solids, the waste itself, after filtration, is considered to be the extract for purposes of determining EP toxicity.

[14] The Extraction Procedure test is designed to simulate the physical processes which occur as a result of disposal of toxic wastes in an actively decomposing municipal landfill which overlies a ground-water aquifer. The test uses an acetic acid leaching medium to simulate active decomposition and thereby identify wastes which are likely to leach toxic contaminants into the ground-water.

[15] EPA has *proposed* regulations to amend the definition of "solid waste" to eliminate the current regulation of all wastes that are "sometimes discarded". 48 *Fed. Reg.* 14472 (April 4, 1983). Under this proposal, whether or not materials fall within the definition of solid waste generally would depend upon how they are actually handled by a particular generator. In addition, subject to certain conditions and exceptions, the proposed regulations would generally exempt materials which are reclaimed from regulation, unless they are stored in a surface impoundment prior to reclamation, or are reclaimed or otherwise processed in a surface impoundment.

[16] Pending statutory amendments may mandate that EPA develop regulations for the burning and blending of fuel derived from hazardous waste or from used oil.

[17] EPA is studying revisions to the small-quantity generator provisions and expects to issue regulations lowering the exemption levels in 1986. Further impetus for an expedited schedule for such changes may result from Congressional action on pending RCRA amendments.

[18] See discussion of § 262.34, *infra,* in Section III.D.

[19] Appendix VII to Part 261 identifies the constituent which was the basis for listing a particular waste as toxic. Appendix III sets out the test methods to use in making the demonstration that a waste sample does not contain the relevant constituent from Appendix VII.

[20] Leaving 1 in. of residue in the container does not qualify it as empty if the methods commonly employed to remove materials from that type of container have not been used. The inch should be measured from the deepest point of the bottom of the container. 47 *Fed. Reg.* 36093 (Aug. 18, 1982).

[21] EPA is considering a reduction in the 1-in./0.3% limit for large-size empty containers. 47 *Fed. Reg.* 36093.

[22] "Generator" is defined as any person "whose act or process produces hazardous waste" or "whose act first causes a hazardous waste to become subject to regulation" (§ 260.10).

[23] The Part 261 regulations contain test protocols for and descriptions of the properties that cause a waste to meet each of the hazardous waste characteristics. A generator may use alternative test methods if they have been approved by EPA as a result of a rulemaking petition (§§ 260.20-.21).

[24] On-site means "the same or geographically contiguous property which may be divided by public or private right-of-way, provided the entrance and exit between the properties is at a cross-roads intersection, and access is by crossing as opposed to going along, the right-of-way. Noncontiguous properties owned by the same person but connected by a right-of-way which he controls and to which the public does not have access, is also considered on-site property" (§ 260.10).

[25] EPA has proposed to exempt from this 90-day time limit accumulation of up to 55 gallons of waste at satellite collection points. 48 *Fed. Reg.* 118 (Jan. 3, 1983).

[26] The definition of "treatment" is set out *supra* in Section II.F.1. "Disposal" is defined as "the discharge, deposit, injection, dumping, spilling, leaking, or placing of . . . hazardous waste into or on any land or water so that such . . . hazardous waste or any constituent thereof may enter the environment or be emitted into the air or discharged into any waters, including ground waters" (§ 260.10).

[27] A manifest is not required for transporting hazardous waste around the site of generation.

[28] The regulations contain different provisions on the distribution of copies of the manifest for shipments of hazardous waste solely by railroad or in bulk by water. (See §§ 262.23(c) and (d), 263.20(e) and (f)).

[29] Different requirements have been established for Exception Reports on international shipments (§ 262.50(c)).

[30] Note that on-site transportation by the generator does not require a manifest.

[31] Compliance, monitoring, and enforcement of the DOT regulations will continue to be performed by DOT. However, EPA has expressly retained the authority to enforce those DOT regulations which are specifically incorporated by reference in Parts 262 and 263 (§ 263.10 (Note)).

[32] Generators, however, may store hazardous waste on-site for 90 days after generation without either a permit or interim status (262.34).

[33] This definition of "commence construction" was adopted from the Prevention of Significant Deterioration (PSD) regulations issued under the Clean Air Act. 45 *Fed. Reg.* 33069.

[34] Facilities which handle wastes listed or identified as hazardous after October 21, 1980 are not required to notify EPA of their activities. Such reporting would be required only if EPA expressly requires notification in conjunction with revisions of the Section 3001 regulations (see Section 3010(a) of RCRA).

[35] A "totally enclosed treatment facility" is "a facility for the treatment of hazardous waste which is directly connected to an industrial production process and which is constructed and operated in a manner which prevents the release of any hazardous waste or any constituent thereof into the environment during treatment." EPA cites as an example a pipe in which waste acid is neutralized (§ 260.10).

[36] An "elementary neutralization unit" is a "tank, container, transport vehicle, or vessel" which is used for neutralizing wastes which are hazardous only because they exhibit the "corrosivity characteristic" or are listed in Subpart D of Part 261 only due to corrosivity (§ 260.10).

[37] Generally, a "wastewater treatment unit" is a "tank", which is part of a wastewater treatment system regulated under Section 402 or 307(b) of the Clean Water Act, *and* which receives and treats or stores an influent wastewater that is a RCRA "hazardous waste", *or* which generates, treats, or stores a wastewater sludge that is a RCRA "hazardous waste" (§ 260.10).

[38] A facility does not have to implement thse security requirements if physical contact with or disturbance of wastes, equipment, or structures on active portions of the facility will not injure unknowing or unauthorized persons or violate the requirements of Part 265 (§ 265.14(a)).

[39] In addition, the regulations require a facility owner or operator to notify EPA 4 weeks prior to receiving the initial hazardous waste shipment from a foreign source (§ 265.12(a)).

[40] "Significant discrepancies in quantity are (1) for bulk waste, variations greater than 10 percent in weight, and (2) for batch waste, any variation in piece count, such as discrepancy in one drum in a truckload. Significant discrepancies in type are obvious differences which can be discovered by inspection or waste analysis, such as waste solvent substituted for waste acid, or toxic constituents not reported on the manifest or shipping paper" (§ 265.72(a)).

[41] The Agency suggests that facilities which accept unmanifested wastes subject to a small quantity exemption either obtain a certification from the generator that the waste is exempt from the manifest requirement or file an unmanifested waste report (§ 265.76 (Note)).

[42] These parameters include chloride, iron, manganese, phenols, sodium, sulfate, pH, specific conductance, total organic carbon, total organic halogen, and additional parameters characterizing the suitability of ground-water for drinking water (§ 265.92(b); Part 265, App. III).

[43] A "disposal facility" is defined as a facility "at which hazardous waste is intentionally placed into or on any land or water, and at which waste will remain after closure" (§ 260.10).

[44] Such a disturbance will only be permitted if the owner or operator can demonstrate that (1) it is necessary to the proposed use of the land and will not increase the potential hazard to human health or the environment; or (2) it is necessary to reduce a threat to human health or the environment (§ 265.117(c)).

[45] The regulations also address the use of state-required financial assurance mechanisms (§ 265.149), and state assumption of responsibility for closure and post-closure costs or for compliance with liability requirements for sudden and non-sudden occurrences (§ 265.150).

[46] As with the interim status standards, compliance with the permanent standards does not preclude the government from bringing suit against a permitted facility under the statutory "imminent hazard" provisions (§ 264.4).

[47] EPA interprets the term "owner" in this context to exclude those who both hold bare legal title solely for the purpose of providing security for a financing agreement and do not exercise any of the effective incidents of ownership or equitable title over the hazardous waste facility. 45 *Fed. Reg.* 77490 (Nov. 10, 1980).

[48] EPA has interpreted this requirement to mean that the signer must be a president, secretary, treasurer, any vice-president of the corporation in charge of a principal business function, or any other person who performs similar policy-making functions for the corporation. 45 *Fed. Reg.* 52149 (Aug. 6, 1980). EPA has proposed and is expected to finalize a provision for variances from the requirement for the owner's signature under certain circumstances. 47 *Fed. Reg.* 32038 (July 23, 1982).

[49] EPA has proposed to amend its regulations to provide that RCRA permits instead be issued for the designated operating life of each facility. 48 *Fed. Reg.* 5872 (Feb. 8, 1983). However, state regulations may be more stringent and impose shorter permit terms. Pending federal amendments to RCRA may limit the permit term to the current maximum of 10 years.

[50] An applicant may request that EPA exercise its discretion to process a RCRA permit application together with applications for other permits for the same facility (§ 124.4(c)(3)).

[51] Denial of requests for modification, revocation and reissuance, or termination are not subject to a hearing or comment period, but an informal appeal for the denial is a prerequisite for judicial review. EPA is required to send a brief statement of the reasons for the denial to the applicant (§ 124.5(b)).

[52] A fact sheet must be prepared for draft permits for major hazardous waste management facilities and draft permits in which there is widespread public interest or which raise major issues. The fact sheet must set forth the significant factual, legal, methodological and policy questions considered in preparing the draft (§ 124.8).

[53] EPA must grant a formal evidentiary hearing only for terminations of a RCRA permit or interim status (§ 124.71(a)).

[54] Any issue not raised or evidence not submitted during the public comment or hearing period on a tentative decision to terminate will not be considered at any subsequent evidentiary hearing unless good cause is shown for the omission (§ 124.76).

[55] Because of certain ambiguities in EPA's evidentiary hearing regulations, companies should consider filing an appeal on a RCRA permit even though a request for an evidentiary hearing has been or will be filed on the permit in a proceeding consolidated with an NPDES permit. Otherwise, the right to judicial review may be lost.

[56] Component A encompasses tanks, containers, waste piles, and surface impoundments. Component B includes incinerators. Component C covers landfills and land treatment facilities.

Chapter 7

THE LEGAL DEVELOPMENT OF REMEDIES FOR TOXIC TORT PLAINTIFFS — COMMON LAW REMEDIES*

Mary Jean Marvin and Kenneth G. Bartlett

TABLE OF CONTENTS

* All footnotes appear at end of chapter.

I. INTRODUCTION

The age of toxic torts appears to be upon us. . . . [P]rivate industry and the government will soon have to defend more and larger claims for compensation for injuries from toxic substances in the workplace, for example, to asbestos, and for exposure to toxic consumer products, for example, to the drug diethylstilbestrol (DES), involving thousands of plaintiffs and billions of dollars in alleged damages. Toxic tort suits based on environmental exposure have been less prominent, perhaps in part because it is more difficult to document long-term exposure to chemicals than it is in the workplace, where they are handled daily, or in drugs, which are ingested regularly.[1]

Reports concerning personal injuries resulting from exposure to toxic substances — so-called "toxic torts"[2] — have been appearing in the news media with increasing frequency during the past several years. The worst industrial accident to date resulted in the mass exposure of the population of Bhopal, India to the chemical methyl isocyanate, used in pesticide manufacturing. More than 2500 people died in December 1984 as the result of a chemical explosion that released a poisonous cloud.[3] The prospect of long latency effects from this disaster looms heavily. Congress, state legislatures, administrative agencies, and state and federal courts have begun to grapple with the complex issues raised by toxic tort litigation. A wide range of legal, scientific, financial, and industrial authorities representing diverse interests have debated whether the existing compensation system should be modified to address perceived inadequacies. "The current issue is whether the underlying problem is sufficiently serious nationally and sufficiently resistant to solution through state common law and statutes to warrant a federal compensation scheme using either shortcuts to tort liability, direct federal compensation, or both."[4]

Toxic tort actions have been regarded as specialized products liability cases which hold manufacturers and sellers legally liable for injuries due to goods that create unreasonable risks of harm.[5] Generally, three elements must be proven in a products liability case: (1) plaintiff was injured by the product; (2) injury occurred because of an unreasonably unsafe or defective product; and (3) the defect or danger existed when the product left the control of defendant.[6]

Toxic tort plaintiffs often face substantial legal, scientific, and economic obstacles in bringing a successful case. Plaintiff may be time-barred from pursuing a claim by the applicable statute of limitations if several years pass before a latent disease is discovered or before plaintiff suspects that the injury is linked to a toxic exposure. It may be difficult to establish with sufficient legal and scientific certainty that an injury was in fact caused by the gradual toxic effects of a substance. This may be complicated by factors such as diet and smoking. In the case of chemical dumps and generic drugs, it may be hard to identify the particular defendant alleged to have caused plaintiff's condition. Perhaps the defendant is insolvent, not still in business, or otherwise unable to finance compensation. Furthermore, plaintiff may be unable to afford the high transactional costs associated with lengthy personal injury litigation, extensive discovery proceedings, and expert witness fees.

Before proceeding in Section III to a discussion of the common law remedies available to toxic tort plaintiffs, this chapter focuses on major developments in mass exposure cases to explore the extent to which toxics litigation has been overwhelming the court system, and to highlight emerging trends in case management.

II. THE MASS EXPOSURE CASES

A. Introduction
In his article, "The Causal Connection in Mass Exposure Cases: A 'Public Law' Vision of the Tort System,"[7] Harvard Law Professor David Rosenberg is foreboding:

The words asbestos, Agent Orange, and Agent White, Three-Mile Island, dioxin, and a string of acronyms — DES, PCB, PPB, and IUD — signify the dangers of mass exposure torts. Because thorough premarket testing of toxic agents is impractical, and because industry has little incentive to investigate and report mass exposure accidents, the number and variety of accidents that have recently come to light leave one with the disturbing sense that many hazards have yet to be created, let alone discovered.[8]

Mass exposure cases have arisen in four broad areas of toxic torts:

1.   The exposure of workers to toxic substances during the course of their employment;
2.   The exposure of U.S. servicemen and women to toxic substances while on military duty;
3.   The exposure of the public to toxic substances resulting from pharmaceutical use;
4.   The exposure of the public to toxic substances in their environment.

B. Routes of Exposure

This section will briefly describe the major litigation resulting from these incidents of exposure.

1. Worker Exposure

The most notable example of worker exposure to date is asbestos. More than 25,000 asbestos-related personal injury lawsuits are pending in both state and federal courts nationwide.[9] The heaviest caseloads occur in areas with government contract shipyards and U.S. Navy shipbuilding facilities.[10] A study of closed asbestos claims by the Rand Institute for Civil Justice indicates that three worker groups have comprised the bulk of this litigation: shipyard workers, asbestos-related factory workers, and insulation workers.[11]

The nation's largest asbestos producer, the Manville Corporation (formerly Johns-Manville Corporation), filed for bankruptcy under the reorganization provision of the Bankruptcy Code (Chapter 11),[12] citing the anticipated costs of defending against and paying asbestos claims. Amatex and Unarco, two other asbestos producers, have also filed for Chapter 11 reorganization. A bankruptcy court has decided that Manville did not act in bad faith when it filed under Chapter 11.[13] This court also held that the "parties in interest" — i.e., parties entitled to the appointment of a representative in Manville's reorganization proceedings — include those persons exposed to asbestos prior to the bankruptcy filing who may later develop asbestos-related diseases.

It should be noted that in early 1983 the Washington Supreme Court ruled that the bankruptcy of Manville and Unarco did not stay litigation against other codefendants, nor were the bankrupt companies considered indispensable parties to that litigation.[14] These issues arose in consolidated cases arising from approximately 400 lawsuits alleging occupational asbestos exposure. Some 30 companies engaged in the manufacture and sale of asbestos products were named as defendants. The holdings in the consolidated cases appear to be the unanimous view of today's courts on these issues.

2. Military Service-Related Exposure

Exposure to radiation and the herbicide Agent Orange are two major examples of the second category of toxic torts, i.e., military service-related exposures. Lawsuits against the U.S. brought by servicemen who were exposed to radiation during atomic testing are barred by *Feres* v. *United States*[15] which denied recovery for injuries suffered by military personnel incident to active duty military service.

Consistent with the *Feres* line of cases, the claim for medical care, in a lawsuit filed against the government by Vietnam veterans exposed to Agent Orange, was dismissed in June 1984.[16]

However, the massive class action suit against seven manufacturers of Agent Orange has resulted in a settlement of $180 million.[17] The chemical company defendants include Dow Chemical Co., Monsanto Co., Diamond Shamrock Corp., Uniroyal Inc., Hercules Inc., Thompson Chemical Co., and T. H. Agriculture & Nutrition Co. Estimated at 2.4 million Vietnam veterans, the class includes:

> . . . all people who were in the Australian, New Zealand or U.S. armed forces at any time from 1961 to 1972 who were injured by exposure to Agent Orange while in or near Vietnam. The class also includes spouses, parents and children of the veterans directly or derivatively injured as a result of the exposure.[18]

U.S. District Chief Judge Jack B. Weinstein gave final approval to the settlement in January 1985 and approved the fund distribution plan in May of that year.[19]

Even though the settlement fund will not adequately compensate the veterans for their injuries, the compromise was considered fair for several reasons. Among these were that trials would take years and the possibility of recovery through trial was greatly diminished because of the inconclusive scientific evidence on the issue of causation, i.e., whether Agent Orange is a significant factor in causing or contributing to cancer and birth defects. The opinion also enumerated other problems with the case including the government contractor defense and "the inability of individual veterans to prove which of seven corporate defendants manufactured the Agent Orange to which they were exposed."[20]

At least several hundred veterans and their families "opted out" of the class action settlement; their claims were dismissed due to the government contractor defense and on other grounds.[20a]

The Agent Orange case is unusual because in late 1983 Judge Weinstein created two classes of the same individual plaintiffs to determine compensatory and punitive damages. Apparently, no plaintiff may opt out of the punitive damages class.

3. Public Exposure via Pharmaceuticals

Perhaps the most well-known mass exposure cases resulting from pharmaceutical use involve diethylstilbestrol (DES) and Bendectin. Widely prescribed as a miscarriage preventive from 1947 to 1971, DES, a synthetic estrogen compound, has been named the causative agent in thousands of lawsuits alleging reproductive system injuries (precancerous and cancerous abnormalties) in the offspring of pregnant women who took the drug. An estimated two million women ingested DES in the higher dosages prescribed to prevent miscarriage.[21]

In the DES cases, plaintiffs must overcome the burden of identifying which of several possible manufacturers produced the generic drug taken. Several innovative courts wrestling with the identification issue have eased this difficult requirement of proof.[22]

In contrast to DES, Bendectin was produced by only one manufacturer, Merrell Dow Pharmaceuticals, and prescribed to pregnant women as an antinauseant. Since 1977 more than 700 cases have been filed which allege that the drug has caused birth defects such as cleft palates and missing limbs in children. Evidence suggests that millions of women took Bendectin in the 27 years it was distributed. Merrell Dow ceased production in 1983. The company entered into a tentative settlement of the multidistrict litigation[23] for $120 million. In October 1984, however, the 6th Circuit Court of Appeals, citing procedural errors, decertified the mandatory class of 700 litigants that was created 4 months earlier by U.S. District Chief Judge Carl B. Rubin for purposes of settlement.[24] This was subsequently tried in Cincinnati before a federal jury in February of 1985. Even though the plaintiffs' evidence was influential, they failed to prove to the jury that Bendectin caused birth defects. While the verdict for Merrell Dow is a big victory, hundreds of cases remain in both state and federal courts.

4. Public Exposure via Contaminated Environment

The public's exposure to environmental contamination has been the subject of considerable attention. Toxic constituents migrating from landfills or otherwise improperly disposed have contaminated air, land, and sources of drinking water in homes surrounding disposal sites. Love Canal (New York), Triana (Alabama), Times Beach (Missouri), Woburn (Massachusetts), Jackson Township (New Jersey), and Silicon Valley (California) are several of the more dramatic examples of personal injuries resulting from unsafe waste disposal to date.

People may be exposed to harmful chemicals in other ways, for example, by the ingestion of lead paint or by the off-gassing of formaldehyde fumes from home insulation. Food products such as milk, fish, bread, vegetables, meat, and poultry can become tainted with contaminants such as PCBs, PBBs, EDB, mercury, and pesticides.

In a recent case,[25] the U.S. was held liable on negligence theories for civilian injuries and cancer caused by atomic testing in Nevada. In May 1984, the district court in Salt Lake City, Utah awarded $2.66 million to 10 victims and their families in a representative group of 24 cases (the "bellwether cases") out of the thousands of people who were exposed to radioactive fallout in the 1950s. The cases were tried without a jury, pursuant to the Federal Tort Claims Act[26] under which the claims were brought.

Ayers v. *Township of Jackson,*[27] one of the first toxic pollution cases, involved 350 members of 97 families who alleged that contamination of their drinking water caused both personal injury and property damage. Plaintiffs' complaint, couched in negligence, strict tort liability, nuisance, trespass, and battery, alleged that defendants — a municipality and a township engineer — were negligent on several counts in operating a 135-acre landfill in Ocean County, New Jersey. The residents claimed that their underground water supply contained acetone, residual aircraft fuel, and benzene and that the contamination caused such harms as emotional distress, enhanced risk of cancer, impaired quality of life, and bodily injury. Plaintiffs also asserted that their exposure to known carcinogens would require future medical surveillance.

On November 16, 1983 an Ocean County Superior Court jury awarded plaintiffs $15.6 million in damages after a 10-week trial.[28] Earlier, the court ruled that claims for enhanced risk of disease were too speculative in that the plaintiffs' medical evidence was insufficient to support a cause of action for actual damages.[29]

The jury's $15.6 million damage award included $8.2 million for medical surveillance, $5.4 million for impairment of the quality of life, $2 million for emotional stress, $92,000 for nuisance damages, and $104,000 for connecting homes to the water system.[30] In June 1985 an appeals court threw out the emotional distress and medical surveillance damages; plaintiffs have appealed to the New Jersey Supreme Court.

Other toxic pollution cases have resulted in settlement. For example, Love Canal claims brought by 1325 residents against Occidental Petroleum Corporation led to a multimillion dollar settlement in late 1983.[31] Even more are still pending.[32] And it is predicted that toxic tort filings will experience steady growth as cases spring up around hazardous waste dumps, particularly Superfund sites.[33]

C. Case Management Trends

Several mechanisms have recently been used to streamline the management of mass exposure tort litigation.[34] Both plaintiff and defendant attorneys attempting to reduce transaction costs in toxics cases are seeking more organized methods with which to efficiently handle mass tort cases. Procedural devices that have been employed include class actions,[35] multidistrict litigation[36] treatment, consolidation[37] or bifurcation of trials, joinder[38] of parties, and discovery management techniques. Special masters, magistrates, and expert panels have also been used to reduce the number of disputed issues and to resolve others in complex cases.

The issue of joinder of similar claims focuses on procedure. While not widely used, the present procedural framework may generally be viewed as being adequate to accommodate the consolidation and trial of similar claims. The importance of this is not to be underestimated since cost has been previously cited as a prime obstacle in seeking recovery. Therefore, to the extent that this cost may be shared by procedurally litigating several similar claims in one forum without sacrificing the merits of individual claims, it is desirable. This desirability is heightened in instances where the elements of causation are identical.

Such cost reduction is presently available by voluntary joinder of parties. Fed. R. Civ. Pro. 20(a) states in part: "All persons may join in one action as plaintiffs if they assert any right to relief jointly, severally, or in the alternative in respect of or arising out of the same transaction, occurrence, or series of transactions or occurrences and if any question of law or fact common to all these persons will arise in this action . . . ". Where individual claims are not within the traditional parameters of joinder, they nonetheless may be consolidated for purposes of trial. Similar consolidation and transfer rules exist for federal actions brought in different districts.[39]

Procedural and substantive requirements have diminished the effectiveness of class actions. Stringent notice requirements and requirement of the $10,000.00 minimal amount in controversy both are substantive barriers that have diminished the prospects of wide utilization of this device. (See *Zahn* v. *International Paper Co.*[40]) It has been the law that each named plaintiff must allege more than $10,000.00 in damages. It has been noted that the use of class actions in mass accident and pollution cases has been the exception and not the rule.[41]

Moreover, the procedural and substantive law of the applicable jurisdiction must be examined to determine the desirability of any of the methods for collective treatment of toxic tort cases. In fact, a special master in the Agent Orange litigation has cautioned that: "Mass tort cases involving hundreds of plaintiffs where transactional costs are prohibitive, where causation is dubious, where exposure is very pervasive, and where the product has a generally recognized social utility"[42] should be removed from the judicial system in favor of administrative approaches.

Section III examines the common law causes of action potentially available to toxic tort plaintiffs. It concludes with a discussion of recurring issues raised in latent injury cases and the emerging trends attempting to address them.

III. THE COMMON LAW OF TOXIC TORTS

A. Causes of Action
1. Introduction

Negligence, strict liability, nuisance, and trespass are the four traditional common law causes of action by which defendants may be held liable for personal injuries. The toxic tort case involving hazardous waste pollution effects typically alleges separate counts of each cause of action. Those torts more closely akin to products liability such as worker exposure to asbestos and pharmaceutical injuries are best suited to actions that lie in negligence, strict liability in tort, and breach of warranty.

It will be seen, however, that existing common law tort actions are inadequate to fully compensate those who have been injured as a result of exposure to harmful substances. A potential plaintiff often has difficulty in proving the elements of each cause of action. Such problems include establishing causation, duty, and damages. Due to these proof problems and various policy considerations, strict liability may be the least burdensome approach,[43] although it is not favored in all jurisdictions.[44]

While there is little precedent in the area of liability of hazardous waste generators and transporters for toxic pollution victims, "[c]ommon law is not static."[45] Modern courts have created precedents, especially in the DES cases, to allow recovery when justice demands.

This section examines the common law causes of actions for personal injuries allegedly caused by toxic substance exposure.

2. Negligence

Simply stated, negligence is "[t]he omission to do something which a reasonable man guided by those ordinary considerations which ordinarily regulate human affairs would do, or the doing of something which a reasonable and prudent man would not do."[46] When defendant fails to exercise reasonable care and injures plaintiff, an action in negligence provides the basis for redress. In the field of products liability, all jurisdictions hold that manufacturers of products owe a duty to consumers; liability for negligence of a manufacturer may be imposed when a defective product is the proximate cause of injury.

The traditional elements that plaintiff must establish to mount a *prima facie* case of negligence are (1) the existence of a duty to conform to a standard of conduct that protects others from unreasonable risks; (2) the defendant's (actor's) failure to conform to that standard; (3) a reasonably close causal connection (proximate cause) between the defendant's conduct and the resulting injury; and (4) the plaintiff's actual loss, injury, or damage.[47] One commentator has noted that in a toxic tort case, plaintiff must prove the identity and source of the injurious substance, its pathway to the plaintiff, and the etiology of the alleged disease.[48]

This section on negligence concludes with an analysis of recent developments regarding shifting burdens of proof and ascertaining liability among multiple defendants.

a. Existence of Duty

A finding of negligence requires the existence of a legal duty — owed by defendant to plaintiff — which has been breached. Absent intent to pollute or injure, negligence is an appropriate basis of liability for harms caused by hazardous substances where proof of failure to exercise reasonable care can be established. Dictum by Judge (later Justice) Cardozo in *Adams* v. *Bullock* defined the duty owed in conducting potentially dangerous activities: "Reasonable care in the use of a destructive agency imports a high degree of vigilance."[49]

Unless there is an applicable statute or regulation governing standard of conduct, the jury will decide the legal standard of due care on a case-by-case basis. To accomplish this, the law imposes a balancing test, taking two main factors into account: (1) the magnitude of risks involved and (2) the existing state of knowledge about the activity at the time of the conduct complained of.[50]

Law presumes that it is unreasonable to expect people to avoid harms that are not reasonably foreseeable.[51] If the risk of injury is foreseeable, however, a jury may decide the risk was reasonable; defendant will not be liable for that risk. A risk may be unreasonable if it is of sufficient magnitude and probability to outweigh the utility of the conduct presenting that risk.[52] Some pharmaceuticals, for example, produce side effects, but they are marketed because these risks are balanced against the benefits of their use. Thus, defendant may not be liable for failure to remove these dangers; liability could be found, nevertheless, for failure to warn of them.

As to the second element in the balancing test, toxic tort plaintiffs may have difficulty in proving that defendant owed them a greater duty than that generally imposed on industry at that time. Nevertheless, courts have applied the rule that conforming to an industry-wide custom will not always suffice to relieve a defendant of taking further precautionary measures to avoid risks of injury where dangerous materials are involved.[53] In products liability cases implicating manufacturers of such dangerous products, a number of courts have held these defendants to be in constructive possession of the skill and knowledge of experts. The court in *Borel* v. *Fibreboard Paper Products Corporation* stated that "[t]he manufacturer's status as expert means that at a minimum he must keep abreast of scientific knowledge, discoveries, and advances and is presumed to know what is imparted thereby."[54]

b. Failure to Conform to Duty

After the existence of a duty owed by defendant to plaintiff has been established, plaintiff next must prove that defendant breached this duty. Several theories are available. One of these consists of introducing direct evidence, usually by means of expert testimony. For example, defendant may have disposed of toxic chemicals in an unreasonably dangerous manner or maintained a disposal site improperly thereby breaching the duty of care owed to nearby residents. A chemical manufacturer might then defend on the ground that due to the lack of scientific knowledge on safe disposal and storage methods for toxic materials, the company's methods were not unreasonable at the time. In the Love Canal case, however, "[t]he unreasonably low standard of Hooker's care in the management and disposal of their chemical wastes is manifest in the equally low standard of their vigilance."[55]

In New Jersey, for example, it is established case law that when a landowner "undertakes to make artificial changes in the land . . . by the storing or depositing of movable material on his land which afterwards is carried by the ordinary force of nature, the owner becomes liable for his negligence if any."[56]

Alternatively, plaintiffs could advance the theory that defendant chemical company was negligent in failing to warn residents of potential hazards: had plaintiffs known of the risks involved they would have acted to avert the danger. In *American Cyanamid Co.* v. *Sparto,* a damages award was affirmed on appeal for a family of truck farmers who used river water to irrigate their land. The manufacturing company dumped chemicals into the river which resulted in stunted growth of the farmers' crops. The court held that if defendant created a risk of injury to plaintiffs that might have been avoided by a warning, then defendant chemical company had the duty to warn, the failure of which constituted negligence.[57]

In a products liability case, defendant manufacturer may also be found negligent for improper labeling, design, testing, or marketing of a product, as well as for failure to warn. Manufacturers have a duty to make a product reasonably safe for its intended use.[58]

After consideration of evidence, the jury will weigh the magnitude of risks involved and the existing standard of knowledge to determine the reasonableness of defendant's conduct and whether the duty was breached.

c. Causation

Plaintiff has the difficult burden of proving a causal connection between defendant's conduct and the actual harm. In other words, causation is the nexus between the harms alleged and the chemicals involved in the environmental condition for which the defendant is responsible.[59] This necessitates establishing proximate (or legal) cause which involves three components: (1) cause-in-fact (i.e., the direct connection between cause and effect); (2) foreseeability of harm as resulting from tortious cause; and (3) demonstrable absence of intervening causes.

i. Cause-in-Fact

Causation-in-fact is a jury question involving determinations of that which has, in fact, occurred. Injured plaintiffs may have a difficult time with the legal causation issue due to the very complex natures of chronic toxicity and carcinogenicity. These afflictions are rarely the result of simple chains of causation; not only because of long latency periods, but also because the tolerance levels for many substances are not fully known; nor are all chemical interactions and biochemical pathways completely understood.[60]

An additional complicating factor is the likelihood that more than one possible source of the disease is present.[61] Indeed, the former President of the American Public Health Association has stated:

> It is only occasionally possible to demonstrate cause and effect between a toxic substance and a specific health condition. An example where this direct relationship exists is asbestosis. In order to have asbestosis, a person must have been exposed to asbestos. Such relationships are a rare phenomenon, however, since most health effects of toxic substances are not specific. Most conditions caused by exposure to toxic substances, as examples, liver cancer, miscarriage, or spontaneous abortion, could also occur in populations not exposed.[62]

In tort law, toxic tort injuries are placed on an equal footing with other injuries. Plaintiffs have the burden of introducing evidence which affords a reasonable basis for the conclusion that the causation element — i.e., how the particular defendant inflicted discrete harms — is satisfied. Where the conclusion is not within the common knowledge of laymen, expert testimony may provide an acceptable and sufficient basis for it.

In a toxic tort case, extensive expert testimony to prove causation will be required because the relationship between chemicals involved and harms alleged is highly technical, beyond the average juror's cognizance. Testimony would have to demonstrate that defendant's behavior more probably than not caused the injuries of which plaintiff complains.[63]

Causation is never proved absolutely; it is established inferentially from a view of the evidence. The *Restatement (Second) of Torts* Section 430 provides that an actor's negligent conduct is a legal cause of harm when that actor's conduct is a substantial factor in bringing about the harm and there is no legal exemption from liability.

Practical and theoretical difficulties are involved in proving causation. The state of current scientific knowledge and the skill of medical diagnosis do not always lead to conclusive legal or scientific evidence of specific causes of a disease. The following observation helps to clarify the situation:

> In law, causation is generally proved by demonstrating that it is more likely than not that X caused Y; in the sciences underlying toxic substances, most notably toxicology and epidemiology, causation is accepted when there is less than a five per cent chance that the association could be accidental. Thus, the rigor required of scientific proof may prevent the generation of information adequate to support legal recoveries.[64]

In toxic tort cases, much of the proof of causation depends on statistical, epidemiological, immunological, and medical evidence relating the substance to particular diseases. Epidemiology correlates the incidence (rate of new cases), prevalence (distribution of cases per area per time period), and determinants of disease in human populations by examining the interrelationship of man and his environment. The following excerpt is instructive:

> Courts have found it difficult to apply this (preponderance-of-the-evidence) standard to the kind of evidence seen in toxic tort litigation, and as a result, have sometimes allowed recovery based on highly suspect evidence, or conversely, have failed adequately to justify the exclusion of evidence. These problems can be overcome, however, if courts apply epidemiologic principles and concepts in conjunction with the traditional standard of proof. Epidemiology is the only generally accepted scientific discipline that deals with the integrated use of statistics and

biological/medical science to identify and establish the cause of human diseases. Its use enables scientific estimation of the percentage of the risk of a disease that is properly attributable to a given factor, such as exposure to an allegedly harmful substance. Thus, use of an epidemiologic standard would provide courts with a rational and consistent means for evaluating evidence of a causal relationship between exposure to a particular factor and the incidence of a disease. . . . [Plaintiff's] burden includes the production of evidence from which the factfinder could reasonably infer that the accused substance "more likely than not" caused the plaintiff's harm. The plaintiff must introduce evidence of both the substance's harmfulness at a given exposure level, and of his exposure to the substance at or above that level.[65]

Thus, bringing a successful toxic tort case requires gathering and presenting a significant amount of scientific evidence such as statistics, epidemiology, and medical testimony. Of course, evidence of exposure is also needed, including information on the relationship of the nature of exposure (in air, water, etc.), the kind of contact or route of entry (by dermal contact, ingestion, inhalation, etc.), and the duration, frequency, and locational proximity of the injured person to the toxic substance.

When and whether all this evidence is available, it becomes apparent not only that expert witnesses may be difficult to secure, but also that the costs of presenting such complex scientific evidence may be quite high. Perhaps this is the reason why so few plaintiffs have decided to litigate their claims to date, preferring to settle whenever possible for frequently nominal sums.

An indication of the extent of expert testimony required in a toxic pollution case has been commented upon:

Toxic waste plaintiffs alleging damages from chemical pollution of their water supply may attempt to produce testimony from hydrologists, geologists, civil engineers, and topographers as to the path, direction, and velocity of the water flow. In addition, bacteriologists or chemists can be called upon to sample and identify the toxic material in the water and to describe its effect. Plaintiffs who claim pollution of their household air through the migration of chemical vapors into their basements, as in Love Canal, may use the testimony of a physicist or industrial hygienist to identify any chemical found in the air and explain its effect. Further expert testimony may come from physicians, experts in industrial medicine, or research veterinarians. If the physician has been the plaintiff's family doctor for a long period of time, his or her chronicle of the plaintiff's medical history could be especially important in eliminating alternative causes of the affliction. The testimony of the industrial medicine specialist can be used to try to reinforce that of the industrial hygienist as to the effects of the contaminant at issue. The research veterinarian may report upon the toxic effects of the chemical on laboratory animals.[66]

ii. Foreseeability of Harm

The second element in the proximate cause formula is foreseeability of the risk that an injurious effect will result from the tortious cause. (The relationship between foreseeability and the exercise of due care is described before.)

iii. Intervening Causes

Finally, proof of causation hinges upon the absence of an intervening cause; i.e., an independent act, event, or agency that produces a result which would not otherwise be reasonably anticipated. Such an event destroys the causal connection between the original wrongdoer and the injury. For example, Hooker Chemical Company's inability to identify another agency that added to the Love Canal dump-site during breaches of the clay cover negates any claim of intervening causes that might exculpate the company from liability.

d. Actual Damage

The fourth element of proof in any negligence case is that actual loss or injury occurred. Cases in the toxic torts field involving chronic exposures and latent injuries rather than acute effects may raise unusual damage problems. Establishing the existence of ordinary damages — for physical injury, medical expenses, loss of earnings, and pain and suffering — is seldom difficult (although the nature of the continuous disablement may arise more frequently in chemical-caused injuries). However, other damages may be hard to quantify, especially those damages for future loss of health. Psychic distress is also difficult to prove. This includes cancerphobia as well as the fear of bearing defective offspring as a result of exposure to substances with carcinogenic, mutagenic, teratogenic, or fetotoxic effects.[67] Injuries often considered too speculative to be compensable are loss of life expectancy and loss of the ability to bear children.

In a toxic pollution case, science may indicate that a group of people residing near a leaking landfill has a particular form of cancer. Some in the group may have contracted that cancer from exposure to certain wastes, but it may be difficult to identify particular individuals affected by the substances given the uncertainty of causation. "Thus, unless a majority of a group of plaintiffs appear to have been injured by the defendant, there is no stirring factual persuasion to grant recovery in any specific case."[68]

Because science is still at the elementary stage of knowing with certainty that exposure to toxic substances can cause or contribute significantly to such insidious diseases as cancer, some individuals may simply be unaware of the possible connection between their illness and chemical exposure. Other factors such as smoking, diet, and the aging process can affect one's physical well-being; each factor alone may be a substantial cause of injury or could interact to produce illness.

Courts, however, are in the process of fashioning new remedies for toxic torts and several approaches in the nature of theories of damages have started to emerge. Some jurisdictions allow recovery for the enhanced risk of contracting a future disease. The measure of damages is based on a discounted standard that reduces the traditional elements of compensation to account for future illness and the percentage likelihood that injury later will become manifest at all.[69] Of course, the enhanced risk must be related to a presently existing condition, to a reasonable degree of scientific certainty. The cost of medical surveillance of exposed communities may be recovered as medical services necessitated by tortious activity.

Where contamination of a drinking water supply has occurred, recovery may be had for impaired quality of life.[70] Additionally, ground-water pollution that constitutes a nuisance may give rise to recovery for emotional distress where plaintiff's property has been damaged or devalued.

3. Strict Liability
a. Abnormally Dangerous Activities

Strict liability is the second common law theory that is potentially available to a toxic tort plaintiff. *Rylands* v. *Fletcher* set forth the original rationale, placing the onus for damages on the defendant who, during a non-natural use of his land, maintains a source of "mischief" which escapes and causes foreseeable harm to the person or property of another:

> We think the true rule of law is that the person who for his own purposes brings on his land and collects and keeps there anything likely to do mischief if it escapes, must keep it at his peril, and if he does not do so is *prima facie* answerable for all the damage which is the natural consequence of its escape.[71]

Section 519 of the *Restatement (Second) of Torts* (1977) states: "[o]ne who carries on an abnormally dangerous activity is subject to liability for harm to the person, land, or chattels of another resulting from the activity, although he has exercised the utmost care to prevent the harm."

Thus, the crucial issue for plaintiff is whether toxic waste disposal and/or the continued maintenance of a disposal site, for example, are abnormally dangerous activities. Courts may define "abnormally dangerous" activities differently.

The *Restatement of Torts* has supplied several tests under which strict liability for hazardous wastes may be imposed. The first *Restatement of Torts* § 520 (1934) imposed liability for "ultrahazardous activities" as follows: (1) An activity is ultrahazardous if it necessarily involves a risk of serious harm to the person, land, or chattels of others which cannot be eliminated by the exercise of the utmost care, and (2) is not a matter of common usage. This has been criticized as going beyond *Rylands* by ignoring the location of the act and its surroundings; furthermore, it falls short of *Rylands* by requiring that the extremely dangerous activity cannot be eliminated even by all possible care.[72]

The *Restatement (Second) of Torts* changed the critical term to "abnormally dangerous" activities. There are several factors to be considered in determining when an activity is "abnormally dangerous":

1. Existence of a high degree of risk of some harm to the person, land, or chattels of others;
2. Likelihood that the harm that results from it will be great;
3. Inability to eliminate the risk by the exercise of reasonable care;
4. Extent to which the activity is not a matter of common usage;
5. Inappropriateness of the activity to the place where it is carried on; and
6. Extent to which its value to the community is outweighed by its dangerous attributes.[73]

A court using this formulation to determine what constitutes an "abnormally dangerous" activity will give dominant consideration to the activity's relation to the totality of circumstances; it must be judged by the relevant societal, economic, and environmental parameters within which it is conducted.[74]

A case that can be analogized to a leaking hazardous waste site was before the Florida Supreme Court in 1975. In *Cities Service Co.* v. *Florida,* the company generated phosphate slimes as a result of mining operations, and deposited the wastes in a settling pond. The state brought suit when the slimes contaminated nearby waterways after breaking through a dam. Trial court gave the state a partial summary judgment which the high court, quoting *Rylands,* affirmed: "Admitting the desirability of (the underlying activity), the rights of adjoining landowners and the interests of the public in our environment require the imposition of a doctrine which places the burden upon the parties whose activity made it possible for damages to occur."[75]

Not all courts have adopted strict liability for abnormally dangerous activities; however, Professor Prosser has noted that "even the jurisdictions which reject *Rylands* v. *Fletcher* by name have accepted and applied the principle of the case under the cloak of various theories. Most frequently . . . the same strict liability is imposed upon defendants under the name of nuisance."[76]

The potential dangers of hazardous waste disposal areas can be analyzed with respect to the blasting cases. In Love Canal, for example, underground interactions of different chemicals (unneutralized and unsegregated) occurred over a period of many years, giving rise to an extremely dangerous situation, even though the method of disposal may have been a legitimate commercial practice at that time. Discussing the blasting cases, Professor Prosser observed:

> [T]he strict liability is entirely a question of the relation of the activity to its surroundings; . . . the use of explosives on an uninhabited mountainside is a matter of negligence only, while anyone who blasts in the center of a large city does so at his peril, and must bear the responsibility for the damage he does, despite all proper care.[77]

Toxic tort plaintiffs alleging strict liability are more likely to succeed if it can be shown that, at the time the defendant buried the wastes, the disposal methods inevitably and necessarily threatened a high degree of harm to others.[78]

What follows are some concluding remarks on the use of Section 520 of the *Restatement (Second) of Torts* in toxic tort cases:

> The extent to which the particular chemical or substance posed an unknown or uncertain health risk at the time of the alleged injury may have a significant influence on the application of § 520 of the Restatement, especially provisions (a) to (c) and (e). That is, the more advanced the state of toxicologic knowledge about a particular substance, the easier it will be for a court to determine whether activities involving that substance are abnormally dangerous.[79]

Thus, "The doctrine of strict liability, if applicable, would eliminate many of the impediments to recovery posed by other common law doctrines . . . Under strict liability, anyone injured by defendants' actions can sue, and plaintiffs need not prove fault."[80]

On the other hand, not all commentators are as optimistic about the likelihood of success in or the advantages of an action based upon strict liability.

> Many courts have been reluctant to extend the doctrine of strict liability for abnormally dangerous activities to the manufacturers and sellers of the products that cause an activity to be abnormally dangerous. In addition, evidence introduced to prove such a theory of recovery may undercut the victim's chances of recovery under other tort theories. This is because proof that an activity is abnormally dangerous requires the plaintiff to show that the risk inherent in the activity cannot be eliminated even with the defendant's utmost care. Such a showing may be inconsistent with the victim's argument that the manufacturer should be liable under a negligence or strict liability theory for improperly testing, designing, or labeling a product. In addition, certain products are unlikely to be considered abnormally dangerous when they are not regulated as toxic pollutants under any federal statute or regulation, are of common usage in the community, and are not under the control of the manufacturers when the potential harm occurs. . . . However, this theory of recovery may be of importance in cases where the user of an abnormally dangerous product provides a financially solvent defendant with "deep pockets" to provide for compensation.[81]

Moreover, it has been noted that the theory may fail if plaintiff is unable to prove proximate cause, or if the statute of limitations has run.[82]

b. Strict Products Liability

Products liability is a rule of law holding manufacturers or suppliers of defective products strictly liable for harm resulting from a defective product.

Section 402A of the *Restatement (Second) of Torts* (1965) has been adopted by two thirds of the states. This section provides "for strict liability for physical harm to users or consumers of a product if the product is sold in a defective condition unreasonably dangerous to the user or consumer or to his property, and (1) the seller is engaged in the business of selling such a product; and (2) the product is expected to and does reach the user or consumer without a substantial change in the condition in which it is sold."[83] Although such defenses as disclaimers, lack of privity, and failure to give notice of breach are generally unavailable in a § 402A action, most states limit recovery to users or consumers of products, thereby precluding recovery by bystanders who have been injured by toxic waste contamination.[84]

In the products liability field, strict liability applies to manufacturing defects, defective design, and inadequate warnings. Plaintiff must prove: (1) the product was defective; (2) the defect existed when the product left defendant's control; and (3) the defect caused injury.

By analogy, perhaps, hazardous wastes may qualify as "products" with respect to "injuries from improperly labeled or safeguarded toxic wastes."[85]

Thus, strict liability of sellers and manufacturers for injuries resulting from defective products that are unreasonably dangerous may be imposed. When safely handled, toxic substances that serve useful purposes are not defective. Nevertheless, these chemicals may become defective if they do not carry proper warnings of their dangerous nature.[86] A number of courts are following the holding of the New Jersey Supreme Court in *Beshada* v. *Johns-Manville Products Corporation,*[87] a case arising from asbestos exposures. *Beshada* abolished the state-of-the-art defense in a failure-to-warn strict products liability case; however, in *Feldman* v. *Lederle Laboratories*[87a] the New Jersey Supreme Court tacitly acknowledged that the *Beshada* holding is limited to asbestos exposure cases. Defendant asbestos manufacturers were thus unable to assert the defense that product dangers were either unknown or unknowable when placed in the stream of commerce. Courts are wrestling with both negligence and strict liability issues in an effort to establish consistent theories of liability in the design defect and failure-to-warn cases.[88]

4. Nuisance
a. Private Nuisance

A private nuisance connotes an interference with plaintiff's use and enjoyment of his land. Private nuisance plaintiffs in a toxic substance discharge case must prove an interference that is (1) unreasonable, intentional, and substantial, (2) reckless or negligent, or (3) an abnormally dangerous activity subject to strict liability.[89] (As discussed elsewhere, a toxic tort plaintiff faces substantial obstacles in making such showings.) A nuisance action may include injunctive relief as well as damages.

In addition, demonstrating that the interference was unreasonable and substantial "requires the balancing of the social utility of defendant's activities . . . against the interests advanced by the plaintiff. This balancing of interest may be difficult where the harm is remote and there is evident utility in the waste disposal."[90]

Other issues include such factors as the surrounding land uses, whether plaintiff has "come to the nuisance", the extent of interference, and the severity of injuries.[91]

It has been suggested, however, that maintenance of disposal sites which discharge substantial quantities of toxic substances could easily constitute a private nuisance.[92]

b. Public Nuisance

"A public nuisance is an interference with the rights of the community at large."[93] It may be both a public and a private one at the same time, and a court will weigh the utility of conduct against the harm. A toxic waste dumpsite would probably qualify as an unreasonable interference with the public's right to good health. The public prosecutor usually brings the action to abate a public nuisance, although it is open to question whether a disposal site that has a government license or permit should be exempt from liability. Intentional or negligent conduct need not be proved in a public nuisance action. At least one court has held that there is absolute liability, regardless of fault, for harms caused by a nuisance resulting from the use of an unreasonably dangerous product or inherently dangerous activity.[94]

"A private individual can sue for damages flowing from a public nuisance only if his loss or injury differs in kind (i.e., special damages) from that suffered by the public at large."[95]

Private remedies are not available under most federal and state citizen suit provisions of pollution control statutes, as recent trends indicate. Even though "[p]rivate actions may be available under public nuisance statutes . . . there is clearly no such private right of action left under the federal common law of nuisance."[96] This principle of law was recently enunciated in *Middlesex County Sewerage Authority* v. *National Sea Clammers Association.*[97]

Some state public nuisance statutes do expressly provide for private rights, however. Eleven states (Alabama, Colorado, Georgia, Indiana, Iowa, Montana, North Dakota, Ohio, Oklahoma, South Dakota, and Washington) have enacted general noncriminal nuisance statutes. At least nine (Arizona, Delaware, Kansas, Nevada, New Jersey, New Mexico, New York, Utah, and Washington) have criminal statutes. Additionally, the following 14 states provide private rights of action for statutory health-related nuisances: Alabama, California, Colorado, Georgia, Idaho, Indiana, Iowa, Oklahoma, South Carolina, South Dakota, Tennessee, Utah, Washington, and Wisconsin. Whether these actions can be used for hazardous waste situations is an unanswered question.[98]

5. Trespass

Another common law theory that is potentially available to a toxic tort plaintiff is trespass. Trespass involves an intentional invasion of plaintiff's exclusive possessory interest in land, unlike nuisance which concerns an interference with the use and enjoyment of land; but, the two causes of action may coexist. For injuries caused by direct entry of tangible objects onto land (e.g., rocks thrown onto property during construction work), classical trespass would have been the traditional action brought. On the other hand, indirect injuries (e.g., from blasting vibrations) would have been remedied in a trespass on the case action. Today, not all courts continue to make this distinction.[99]

In toxic tort cases, proving intent is perhaps the most difficult element of a case in trespass. Such proof would be insufficient in a situation where the defendant chemical company used disposal or burial methods which created only an increased risk of interference with the person or property of another. It has been suggested that if plaintiff fails to establish causation "it will be nearly impossible to prove that the defendant knew that the harm was substantially certain to follow."[100]

Additionally, it has been held that even though defendant may have intentionally released pollutants onto his land, no liability will ensue absent knowledge that conditions such as subterranean currents would carry the toxic chemicals to plaintiff's property.[101] Nevertheless, a finding of intentional conduct may be based on evidence that the plaintiff has given defendant notice of the contamination, or that defendant is on constructive notice of the harms likely caused by his operations.[102] The requisite intent is found "when a reasonable person in the defendant's position would believe that a particular result was substantially certain to follow."[103]

Various commentators have described the advantages of the trespass theory in toxic waste litigation. Cited are (1) longer statutes of limitation than either nuisance or negligence — each time a chemical is released onto plaintiff's property, a new trespass occurs; thus, the statute begins to run anew in a continuing trespass; and (2) no need for judicial balancing required in nuisance cases where plaintiff must have suffered an unreasonable and substantial injury; in trespass, "even *de minimis* injuries may result in the award of damages."[104] Note, however, that courts may engage in the balancing test upon allegations of an abnormally dangerous activity.

Since most courts have abrogated the earlier doctrine that no trespass occurred unless the invasion was by something tangible and visible to the naked eye, trespass may be available in pollution cases where the invasion is by invisible or microscopic chemical particles.[105]

Professor Grad suggests that a recurring trespass may be treated as a nuisance in an air pollution case, but trespass will lie where a "single, isolated . . . unauthorized dumping of toxic wastes on another's land" causes personal and/or property damage. Moreover, in hazardous waste cases, "trespass is likely to involve runoff of liquid wastes."[106]

Because of the stringent proof requirements on the intent issue, and the necessity of proving direct invasion of landowner's property, trespass is seldom the avenue of recovery for plaintiffs injured by environmental pollution. It may be possible to mount a good case of trespass in certain specific situations.

Other causes of action, such as battery, malpractice, fraud, and misrepresentation, may be appropriate. Negligence per se will lie where plaintiff can establish (1) a statutory violation, (2) membership in that class of persons designed to be protected by the statute, and (3) resulting harm that was intended to be protected against by the statute. The theory is that the violation of a statute enacted to safeguard persons and property is a negligent act imposing liability when that negligence proximately causes injury (*Springer* v. *Joseph Schlitz Brewing Company*).[107]

B. Recurring Issues in Toxics Cases
1. Shifting Burdens and Apportioning Damages

Proving that a particular defendant caused plaintiff's injury may be difficult under traditional principles. For example, there may be several hundred manufacturers of a generic drug such as DES. Or, many defendants may be associated with a hazardous waste site case. The most common being waste generators, haulers, site owners and operators, construction firms, and real estate developers.

Several possibilities for shifting burdens of proof and apportioning damages among these potential defendants have been explored by modern courts. The most significant development is the expansion of concerted action liability.[108] This theory has evolved into market share[109] liability, alternative[110] liability, and enterprise[111] liability.

In a concerted action case each defendant will be jointly and severally liable for plaintiff's injury, provided that plaintiff can show defendants acted in concert in furtherance of a common design or plan. This theory may apply to a waste site case or to a toxic tort brought as a products type of action. A jurisdiction allowing concerted action liability will let the case go to the jury even though plaintiff cannot establish which defendant actually caused the harm; thus, this theory is an exception to the rule. In *Bichler* v. *Eli Lilly & Co.,* the New York Court of Appeals ruled that even though plaintiff, a "DES daughter", failed to establish that defendant produced the drug actually ingested by her mother, liability was based on the defendant acting in concert with other drug manufacturers.[112]

A second concept — although not widely used — is enterprise liability, applicable to situations in which there is more than one manufacturer in a centralized industry. In *Hall* v. *E. I. duPont de Nemours & Co.,* the six corporations that made virtually all blasting caps in the U.S. were liable for damages which were pro-rated according to each defendant's share of the market at the time of injury.[113] This theory differs from market share liability first upheld by the California Supreme Court in *Sindell* v. *Abbott Laboratories. Sindell* held that a DES plaintiff need not join all manufacturers (unlike enterprise liability) as long as a substantial percentage of the market share is represented by the defendants whom plaintiff has joined. The burden then shifts to defendants to establish their liability.[114]

Accepting a theory of modified alternative liability, the Michigan Supreme Court allowed DES plaintiffs to proceed without establishing the identity of the manufacturer who actually produced the drug taken by plaintiffs' mother. Defendants had the burden of proving that their product was not responsible. (*Abel* v. *Eli Lilly.*[115]) In another DES case, *Collins* v. *Eli Lilly,* the Wisconsin Supreme Court rejected the market share concept.[116] However, the court ruled that defendants had to prove they were not liable; plaintiffs could claim damages from any one defendant.

2. Statute of Limitations

Simply stated, the statute of limitations period is the time in which an injured party must file suit; it is legislatively determined and understandably reflects a broad social policy against the prosecution of stale claims. An examination of the time period itself reveals that it is surprisingly uniform. The focus of inquiry has been on when this period of time begins to run. With latency periods often exceeding a decade or more, most individuals will be time-barred before any ascertainable injury is evident if this period is construed to commence at the time of the exposure itself.

Most states have adopted the so-called "Discovery Rule", that is, that the statutory period begins to run from the date of discovery of the injury or when the individual reasonably should have discovered the injury. Conceptually, the tortious act is viewed as not being "complete" until the injured party discovers the injury. This gradually evolving rule may be viewed as a middle ground position, evolving both through legislation and by judicial action.[117] A substantial minority of states postpone the statutory period from commencing until plaintiff is actually aware he has a cause of action. Other states allow this period to be either the time at which the injury is capable of ascertainment, or when plaintiff knows (or in the exercise of reasonable judgment should have known) that he has suffered a disease or injury.

This principle has traditionally served as a substantial obstacle in defeating claims brought by plaintiffs seeking compensation for exposure to toxic pollutants. Absent a statute, the time of accrual is decided by the court.[118]

Noted authors in the environmental law field differ as to which event or action causes the statute to run. Basically, though, the writers agree that courts choose variations of the traditional rule (when the negligent act or wrongful conduct occurs, i.e., first exposure to the substance), or the discovery rule (when plaintiff discovers the injury).

Unlike traditional products liability cases, where injury soon follows tortious act, toxic substance disease cases present the dilemma that plaintiff's case may be time-barred by the running of the statute before he or she is even aware of the existence of a latent disease. Some courts have acknowledged this problem and have either adopted the discovery rule or have interpreted "ambiguous" statutes in order to achieve more equitable results. For example, 54 of 91 personal injury claims brought by Love Canal residents were dismissed. Suits had been filed more than 3 years from the date of exposure, thus time-barred by the New York statute.[119]

3. Causation

It is evident that the unequal economic status of injured plaintiff and corporate defendant often favors the defendant. Plaintiffs may simply lack sufficient funds to litigate. Several remedial measures have been suggested to alleviate the seemingly insurmountable burden of establishing causation.

One possibility is the continued reliance on an expansion of the existing system of expert witnesses who are appointed by the court in appropriate cases.

Second, there should be greater reliance on government data relating to causation of pollution-induced diseases.[120] Industry data on file with regulatory agencies may offer indications of toxic chemical effects. For example, the Toxic Substances Control Act (TSCA)[121] requires that companies must give EPA information on whether a particular chemical product presents a substantial risk of injury to health or to the environment.[122] The case of *Continental Oil Co.* v. *Federal Power Commission*[123] stands for the proposition that the Freedom of Information Act, in conjunction with agencies' view that safety test data should not be given trade secret protection, may provide plaintiffs with access to data despite industry's claims of confidentiality. Additionally, transcripts of congressional hearings plus plaintiffs' discovery techniques can provide further information.

Other sources of government data on environmentally related diseases include: the Federal Insecticide, Fungicide, and Rodenticide Act (FIFRA); the Consumer Product Safety Act; the Federal Food, Drug, and Cosmetic Act; the Occupational Safety and Health Act (OSHA); the Federal Water Pollution Control Act (FWPCA); the Federal Safe Drinking Water Act (FSDWA); the Atomic Energy Act; the Resource Conservation and Recovery Act (RCRA); the Clean Air Act (CAA); and the Comprehensive Environmental Response, Compensation, and Liability Act ("Superfund") (CER-CLA).[124]

The third remedial measure to be considered is the reliance on (rebuttable) presumptions in connecting certain exposures to particular diseases. In general, presumptions on the work-related nature of diseases and those which link specific kinds of exposures to specific recurring illnesses have been upheld in state workmen's compensation cases that deal with black lung diseases. The U.S. Supreme Court has upheld this presumption that links work exposure in coal mines to the disease in *Usery* v. *Turner Elkhorn Mining Co.*, under the Federal Black Lung Benefits Act of 1972.[125]

4. Insurance Theories

Another recurring issue in toxic tort litigation is of significance to both plaintiffs and defendants. The Comprehensive General Liability Policy is the type of insurance designed to cover personal injuries allegedly caused by a hazardous substance. Liability often revolves around determining the "trigger of coverage" question. Several theories for triggers are being developed which construe the terms "occurrence" and "bodily injury" with respect to latent diseases. The theory of "injurious exposure" as applied in *Insurance Company of North America* v. *Forty-eight Insulations, Inc.*[126] provides that all insurers on the risk during the years of claimant's asbestos exposure are liable for injury on a *pro rata* basis.

On the other hand, the "manifestation" theory defines "injury" as clinically evident injury, thus triggering coverage when a disease is "reasonably capable of diagnosis".[127] In another view, the "triple-trigger" theory, any insurer providing coverage between the initial exposure and onset of disease is liable for indemnification.[128] The "Actual injury" theory bases occurrence of injury upon scientific or medical evidence on each claim.[129]

Finally, the "multi-trigger" theory interprets ambiguous insurance policies to effectuate the reasonable expectation of the parties. In a case arising from the DES litigation, it was held that Eli Lilly, a DES manufacturer, could reasonably have expected its insurance policies to cover all future liability for DES injuries (*Eli Lilly & Co.* v. *Home Insurance Co.*[130]). The company, however, would have to select and apply only one policy limit to each injury.

IV. CONCLUSION

This discussion has focused on the growing threat of hazardous waste contamination. Monetary damages for personal and property losses resulting from exposure to toxic chemicals are difficult to recover in private "toxic tort" suits at common law. A survey of existing provisions for personal injury compensation in the federal statutes will be discussed in the following chapter.

The traditional common law theories potentially available to an injured plaintiff are generally considered inadequate to remedy the harm caused. Elements of each cause of action (whether in negligence, strict liability, nuisance, or trespass) may be extremely burdensome — even impossible — to establish with the requisite degrees of legal certainty. Under a negligence theory, plaintiff may have difficulty in proving a *prima facie* case, and also may be subject to the affirmative defenses of contributory negligence, statute of limitations, and indemnification by a third party. Obstacles to recovery in strict liability include statute of limitations, proof of proximate cause, and showing that disposal of toxic wastes should be considered an ''abnormally dangerous'' activity. In a nuisance case, plaintiff must demonstrate the unreasonableness of defendant's conduct; it may even be necessary to prove intentional, negligent, or abnormally dangerous action on the part of defendant. Finally, where a trespass is alleged, the intent requirement is the most difficult to establish.

Thus, problems of causation, duty, and damages are among the recurring factors in toxic tort cases. Indeed, scientific and evidentiary uncertainties often provide the rationale for courts to sidestep issues (with denial of recovery a likely consequence).

Statutes of limitations may effectively bar suit before a plaintiff is aware of a latent disease. However, it is encouraging to note that there is a trend among courts to adopt the discovery rule (or variations thereof) in the interest of fairness with a view toward compensating victims of hazardous waste contamination.

EPA regulations requiring owners and operators of toxic waste sites to carry liability insurance for bodily injury and property damage to third parties resulting from facility operations, should help alleviate some of the suffering and harm caused by pollution. This, of course, remains to be seen, since potential plaintiffs still may be unable to litigate due to the time and expense involved in maintaining toxic tort actions. Unless an appropriate case management plan is utilized or information is shared among litigation groups, it is certain that difficult legal, scientific, and economic burdens must be overcome.

ADDENDUM

In this rapidly expanding and evolving area of tort law it is a challenge to keep abreast of current developments. (To cite one example, a number of cigarette product liability cases are being filed around the country (see *Litigation Monitor*, Legal Times, Feb. 3, 1986, at 10.)) It is the intent of this addendum to acquaint the reader with the most noteworthy of those developments which did not lend themselves to discussion in the text. Three topics will be highlighted: Tort Reform, Responses to Bhopal, and Civil Racketeering Litigation.

Tort Reform — As alluded to in footnote 4, (see reference to the ABA Special Committee on the Tort Liability System), there is a growing liability insurance crisis facing this country. Several groups have been formed to study the issue and make recommendations on reforming the present system, which has been deemed inadequate by representatives of nearly all parties involved in these cases. For example, the Reagan Administration task force on liability law reform, the Tort Policy Working Group, is expected to recommend that states take responsibility for reform by adopting a model state reform code, which has yet to be drafted by the government. Suggestions for this proposed code include limits on punitive damages and attorneys' fees, a limit on awards for pain and suffering and other noneconomic damages, and a restriction on the use of the joint and several liability standards (where one responsible party may be liable for all damages due from other responsible parties), to a system in which each party pays its fair share. (See, e.g., *Tort Reform*, Nat'l L. J., Dec. 2, 1985, at 3; *Current Developments*, Env't Rep. (BNA), Dec. 20, 1985, at 1631; and *Liability Law Reform: A Job For The States*, Chem. Week, Feb. 5, 1986, at 40. In addition, the Rand Corporation's Institute for Civil Justice has recommended that a national commission be established to address mass toxic tort litigation, based on its study of the asbestos cases now swamping the court system. (See *Rand Urges Tort Commission*, Nat'l L. J., Dec. 16, 1985, at 13.) More recently, a new group has been formed to analyze the system, the business-oriented American Tort Reform Association. (See *New Group Enters Tort-Reform Arena*, Nat'l L. J., Feb. 3, 1986, at 3.) For a comprehensive discussion of both causation and reform issues and recommendations, the reader is encouraged to consult the Institute for Health Policy Analysis Final Conference Report on *Causation and Financial Compensation* (Georgetown University Medical School, 1986).

Responses to Bhopal — On the federal level, the U.S. Environmental Protection Agency has developed a program addressing accidental releases of acutely toxic chemicals in response to the Bhopal, India tragedy of December 1984. The goals of this voluntary preparedness program are to stimulate state and local development of emergency response plans dealing with chemical accidents and to increase community awareness of potential hazards. (See 50 Fed. Reg. 51451, Dec. 17, 1985.) On the state level, New Jersey Governor Thomas H. Kean signed into law that state's new Toxic Catastrophe Prevention Act (A 4145) on January 8, 1986. This has been heralded as the nation's first piece of legislation requiring chemical plants to have plans for avoiding toxic chemical releases. (See *Jersey Acts to Curb Toxic Accidents*, N.Y. Times, Jan. 9, 1986, at B2.) Many other states and local communities are now engaged in examining existing plans to deal with chemical disasters, proposing newer, more responsive approaches where needed. It should also be noted that federal legislation (such as community right-to-know provisions) has been introduced, for example, in the current debate on reauthorization of the Superfund law; these amendments are in direct response to Bhopal.

With respect to personal injury suits brought by the residents of Bhopal (note that other cases have been filed on behalf of Union Carbide shareholders), the Judicial Panel on Multidistrict Litigation transferred all federal cases to the Southern District of New York in February 1985. (*In re Union Carbide Corp.*, MDL 626, S.D.N.Y.) At this writing the proper forum for the litigation has not been determined. (See, e.g., *Negotiations, No Ruling*, Chem. Week, Feb. 26, 1986, at 14; Goodrich, *Bhopal 'Solution' Not Without Flaws*, Nat'l L. J., Mar. 17, 1986, at 13; *Litigation Monitor*, Legal Times, Feb. 3, 1986, at 9.)

Civil Racketeering Litigation — Plaintiffs in hazardous waste cases have recently begun to allege violations of the Racketeer Influenced and Corrupt Organizations Act (RICO), 18 U.S.C.A. Sec. 1961 *et seq.*, in their claims for damages. This trend is expected to continue in light of the U.S. Supreme Court decision in *Sedima S.P.R.L. v. Imrex Co., Inc.*, 105 S. Ct. 3275 (1985), which invalidated many of the limitations on racketeering lawsuits that had been imposed by courts. For example, in *Standard Equipment, Inc. v. Boeing Co.*, No. C84-1129M (W.D. Wash. Feb. 5, 1985), a claim for treble damages was successful. Plaintiff, the owner of property adjoining a hazardous waste dumpsite, allegedly suffered ground-water contamination. Contending that the disposal site was illegal and that mail fraud had occurred in the permit application process, plaintiff cited a violation of section 1962(c), which prohibits conduct of an enterprise "through a pattern of racketeering activity" (i.e., allegedly fraudulent permit applications).

A more recent case, however, brought in Massachusetts, held that conventional claims for personal injuries may not be brought under RICO as injuries to property. in *Moore v. Eli Lilly & Co.* 626 F. Supp. 365 (D. Mass. 1986), plaintiffs attempted to add RICO violations to their complaint, alleging that defendant fraudulently misrepresented facts related to the defective nature of the drug Oraflex. The court ruled that plaintiffs' injuries (diminution in the husband's estate and loss of consortium allegedly suffered by his wife) were not injuries to property but conventional claims for personal injuries. It will be interesting to track developments in this area.

FOOTNOTES

[1] Recovery for Exposure to Hazardous Substances: The Superfund § 301(e) Study and Beyond, 14 Env'tl. L. Rep. (ELR) 10098, 10098 (1984) (hereinafter cited as ELR § 301(e) Article). (Reprinted in this article are the presentations (concerning toxic victim compensation issues) delivered at the Twelfth Annual ABA Standing Committee on Environmental Law, Conference on the Environment, May 6-7, 1983).

[2] Paul Rheingold's article, *Civil Causes of Action for Lung Damage Due to Pollution of Urban Atmosphere,* introduced the phrase "toxic torts". 33 Brooklyn L. Rev. 17 (1966); *Toxic Torts,* Rheingold, P., Landau, N., and Canavan, R., Eds., 1977. Although the term "toxic torts" has achieved some currency of usage, it is not a recognized cause of action. In 1977, the American Trial Lawyers Association published *Toxic Torts: Tort Actions for Cancer and Lung Disease Due to Environmental Pollution,* thus indicative of the acceptance that this term has enjoyed in the legal community. It is descriptive of a new and important category of injuries and situations created by the exposure to hazardous substances. As used in this chapter toxic torts are lawsuits brought by plaintiffs seeking compensation for personal injuries allegedly caused by exposure to substances that increase the risk of contracting a serious disease. Generally, the disease will not become manifest prior to a period of latency (i.e., an insidious disease); toxic torts involve chronic rather than acute effects of occupational or environmental exposure to toxic agents, or pharmaceutical use.

[3] See, e.g., *American Lawyers Flock to Bhopal,* Washington Post, Dec. 12, 1984, at A1, col. 2; *Death Toll Reaches 2500 From Bhopal Gas Leak,* 5 Hazardous Materials Intelligence Rep. 1 (No. 49, Dec. 14, 1984); *U.S. Lawyers Court Disaster in Bhopal,* Nat'l. L. J., Dec. 31, 1984, at 3, col. 2; *The Bhopal Disaster: How It Happened,* N.Y. Times, Jan. 28, 1985, at A1, col. 1 (first in a series of articles). See generally, *Litigation Monitor,* Legal Times, Feb. 3, 1986, at 9.

[4] *ELR § 301(e) Article, supra* note 1, at 10098. A recent 1,000 page report by the ABA Special Committee on the Tort Liability System, entitled *Towards a Jurisprudence of Injury: The Continuing Creation of a System of Substantive Justice in American Tort Law,* contains suggestions on improving the administration of justice. In addressing issues raised by toxic tort cases the report endorsed the "discovery rule" (see *infra* note 117 and accompanying text) cautioned against passage of a federal products liability law, and recommended creating an agency to collect injury data on a nationwide basis. *ABA Report Suggests Changes in Tort System,* Nat'l L.J., Dec. 24, 1984, at 6, col. 1.

Since 1977 Congress has considered a variety of bills on liability for toxic substance exposure, covering both occupational and nonoccupational claims. See, e.g., H.R. 9616, 95th Cong., 1st Sess. (1977); H.R. 2740, 3797, 3798, 5074, 5290, 5291, 6931, S. 1480, 96th Cong., 1st Sess. (1979); in the 98th Cong., see S. 917, 921 (for atomic bomb radiation victims), 945, 946, H.R. 2330, 2482, 2582, 4303, 4813, 5175, 5640 and 5735 (for occupational disease victims). More recent action on federal victim compensation legislation occurred when the U.S. Senate deleted from proposed CERCLA amendments a demonstration victim assistance program (see 131 Cong. Rec. S11998, daily ed. Sept. 24, 1985). Then, in late December 1985, legislation was introduced (S. 1999) that would create a federal products liability law coupled with a victim compensation plan in response to the insurance crisis. State legislation includes N.J. Stat. Ann., 58:10 - 23.11 - 23.11Z, Fla. Stat. Ann. Sec. 376.011-376.21, and Cal. Health and Safety Code, Div. 20, ch. 6.8, Sec. 25300.

Proposals to improve the toxic injury compensation system have been developed by a number of legal commentators. See, e.g., *Injuries and Damages from Hazardous Wastes — Analysis and Improvement of Legal Remedies,* A Report to Congress in Compliance with Section 301(e) of the Comprehensive Environmental Response, Compensation, and Liability Act of 1980 (Pub. L. No. 96-510) by the "Superfund Section 301(e) Study Group," Serial No. 97-12, Pts. 1 and 2, printed for the use of the Senate Comm. on Env'tl and Pub. Works, 97th Cong. 2d Sess. (Comm. Print 1982) (hereinafter cited as the *§ 301(e) Report);* J. Trauberman, *Statutory Reform of "Toxic Torts": Relieving Legal, Scientific, and Economic Burdens on the Chemical Victim* (Env'tl L. Inst. 1983), *reprinted* at 7 Harv. Env'tl L. Rev. 177 (1983) (hereinafter cited as Trauberman); Soble, *A Proposal for the Administrative Compensation of Victims of Toxic Substances Pollution: A Model Act,* 14 Harv. J. Legis. 683 (1977). See also *Six Case Studies of Compensation for Toxic Substances Pollution: Alabama, California, Michigan, Missouri, New Jersey, and Texas,* Serial No. 96-13, A report prepared by the Env'tl L. Inst. with the Cong. Res. Serv. printed for the use of the Senate Comm. on Env't and Pub. Works, 96th Cong., 2d Sess. (Comm. Print 1980) (hereinafter cited as *Six Case Studies);* Gasch and Light, *Evolving Legal Remedies for Toxic Torts:* An Overview, 4 Legal Notes and Viewpoints Q. 1, 22 n. 2 (Aug. 1984) (citing a host of articles on victims compensation). In addition, the Environmental Law Institute (ELI) has recently completed two empirical studies of personal injury cases in Connecticut and Massachusetts courts. At the request of the U.S. Environmental Protection Agency, ELI compared a sampling of resolved Connecticut toxic tort cases brought in state court with a similar number of major, resolved torts not involving latent injuries. Results indicated that: (1) of the 18 resolved toxics cases studied, only 8 resulted in a judicial or workers' compensation award or settlement, compared with 17 out of 18 other torts examined. (The remainder were discontinued actions.) (2) Toxic tort plaintiffs do not recover as frequently as other tort plaintiffs; but once a toxic tort recovery is obtained, it is larger than the other tort recovery. (3) As a percentage of compensation obtained, toxic tort plaintiffs spent 54% in legal costs, compared with 42% spent by other tort plaintiffs. (4) Connecticut state courts appeared to spend more time and resources in resolving toxics cases than other tort cases.

The Massachusetts Executive Office of Environmental Affairs requested ELI to examine the toxics cases in Massachusetts. The following results obtained: (1) There are relatively few cases filed in state and federal courts in Massachusetts that allege hazardous waste site exposures. (2) Sixty-two toxic tort cases (39 occupational exposures; 23 nonoccupational exposures to pharmaceuticals and consumer products) were resolved in federal court from 1980-83 out of a total of 835 closed nonvehicular personal injury cases examined. (3) A total of 1583 toxics cases were filed from 1980-83 in federal court out of a total of 2074 nonvehicular personal injury actions still pending. 1534 of these (or 74% of the total pending caseload) allege asbestos-related injuries. Of the remaining 49 toxics cases, 19 involved occupational exposures and 30 were nonoccupational, including a hazardous waste site case. (4) The average recovery in all closed toxic tort cases examined was only about 2.8% of the recovery in the random sample of other tort cases used for comparative purposes. As a percentage of recoveries, toxic tort litigation costs were strikingly higher than those in other tort cases; toxics plaintiffs spent more in litigation costs than they recovered in court. (5) Nearly 60% of the toxics cases were settled compared to 75% of the other torts. The results of this study are thus not inconsistent with the notion that persons exposed to chemicals have a more difficult time proving their cases than other personal injury claimants.

[5]   See, e.g., McGovern, *Management of Multiparty Toxic Tort Litigation: Case Law and Trends Affecting Case Management*, 19 Forum 1, 1 (1983) (hereinafter cited as McGovern); Bartlett, *The Legal Development of a Viable Remedy for Toxic Pollution Victims*, 10 Barrister 41, 41 (Fall 1983).

[6]   W. Prosser, *Handbook of the Law of Torts* 671-72, 4th ed. (1971) (hereinafter cited as Prosser).

[7]   97 Harv. L. Rev. 849 (1984) (hereinafter cited as Rosenberg).

[8]   *Id.,* at 853-54. See generally, *A Problem That Cannot Be Buried,* Time, Oct. 14, 1985, at 76.

[9]   See, e.g., *id.,* at 852 n. 4; Black and Lilienfeld, *Epidemiologic Proof in Toxic Tort Litigation,* 52 Fordham L. Rev. 732,733 n. 2 (1984) (hereinafter cited as Black and Lilienfeld); Nat'l. Center for State Courts, *Judicial Administration Working Group on Asbestos Litigation: Final Report with Recommendations* 1 (1984) (hereinafter cited as *Working Group Report); Special Project: An Analysis of the Legal, Social and Political Issues Raised by Asbestos Litigation,* 36 Vand. L. Rev. 573, 581 n. 22 (1983).

[10]   *Working Group Report, supra* note 9, at 1.

[11]   Kakalik, Ebener, Felstiner, Haggstrom, and Shanley, Variation in Asbestos Litigation Compensation and Expenses at viii, Rep. No. R-3132 - ICJ (Rand Corp., Inst. for Civil Just. 1984). (This report analyzes characteristics of 513 closed asbestos-injury cases in a random sample of claims and explains variations in compensation and expenses. See also Rand Corp., *Costs of Asbestos Litigation,* Rep. No. R-3071 - ICJ (1983).)

[12]   U.S.C.A. §§ 1101 *et seq.,* as enacted by the Bankruptcy Reform Act of 1978 (Act of Nov. 6, 1978, Pub. L. No. 95-598, 92 Stat. 2549).

[13]   *In re Johns-Manville Corp.,* No. 82 B 11656, 52 U.S.L.W. 2438 (U.S. Bankr. Ct., SNY Jan. 23, 1984); cf., Note, *Manville Corporation and the "Good Faith" Standard for Reorganization Under the Bankruptcy Code,* 14 Toledo L. Rev. 1467, 1506 (1983) (concluding that Manville did not act in good faith thus precluding relief under Chapter 11). See also note, *Tort Creditor Priority in the Secured Credit System: Asbestos Times, the Worst of Times,* 36 Stan. L. Rev. 1045 (1984).

[14]   *In re Stay of Proceeding Against Defendants Johns-Manville Corp. and Unarco Industries,* Inc., 51 U.S.L.W. 2624, 11 Prod. Safety & Liab. Rep. (BNA) 251 (Wash. Mar. 17, 1983).

[15]   340 U.S. 135 (1950). For a related article see *Members of Service Suing the U.S.: Is the Feres Doctrine Crumbling?,* Nat'l. L. J., Mar. 10, 1986, at 34, col. 2.

[16]   Ryan v. Public Health Service, CV-84-2237 (EDNY, June 4, 1984). See also Washington Post, June 5, 1984, at A6, col. 5; Chem. Regulation Rep. (BNA), June 8, 1984, at 302-03.

[17]   *In re Agent Orange Product Liability Litigation,* MDL 381, 13 Mil. L. Rep. (Pub. L. Educ. Inst.) 2076, Memorandum and Order on Attorneys Fees and Final Judgment (EDNY, Jan. 7, 1985). See also *In re Agent Orange Product Liability Litigation,* MDL 381, 597 F. Supp. 740, 12 Mil. L. Rep. (Pub. L. Educ. Inst.) 2692 (EDNY 1984) (Preliminary Memorandum and Order on Settlement).

[18]   Birnbaum and Wrubel, *Agent Orange Class Certification and Industry Wide Liability for DES,* Nat'l. L.J., Feb. 27, 1984, at 38, col. 1, 39, col. 1; 597 F. Supp. 740, 756 (EDNY 1984).

[19]   See *Agent Orange Pact is OK'd by Judge,* Nat'l. L.J., Oct. 8, 1984 at 3, col. 2.

[20]   *Id.,* at 22, col. 3.

[20a]   See *Litigation Monitor,* Legal Times, Feb. 3, 1986 at 10.

[21]   Some estimates are as high as six million women. Johnson and Dowie, *Revenge of a DES Son,* 8 Mother Jones 33 (Feb./Mar. 1982). See also Note, *Emotional Distress Damages for Cancerphobia: A Case for the DES Daughter,* 14 Pac. L. J. 1215, 1219 n. 35 and accompanying text (1983) (hereinafter cited as

Note, *Emotional Distress*); Endress and Sozio, *Market Share Liability: A One Theory Approach Beyond DES,* 1983 Det. C. L. Rev. 1 (1983); Notes and Comments, *Bichler v. Lilly: Applying Concerted Action to the DES Cases,* 3 Pace. L. Rev. 83 (1982).

[22] See, e.g., ELR § 301(e) Article, *supra* note 1, at 10099-100.

[23] *In re Richardson-Merrell Bendectin Products Liability Litigation,* MDL 486 (No. 84-3710, S.D. Ohio 1984). See also *Dow Establishes Fund to Settle Bendectin Suits,* Chem. Eng. News, July 23, 1984, at 7; *Bendectin Pact Creating Furor,* Nat'l L.J., July 30, 1984, at 1, col. 1 (this article contains a report about settlement of the remaining lawsuits over the sedative Thalidomide, at 30).

[24] *In re Bendectin Products Liability Litigation,* 749 F. 2d 300 (6th Cir. 1984); see also *Confusion Reigns Over Bendectin,* Nat'l L.J., Nov. 12, 1984, at 3, col. 1.

[25] *Allen v. United States,* 588 F. Supp. 247 (D. Utah 1984).

[26] 28 U.S.C.A. §§ 2671-80 (1976).

[27] 189 N.J. Super. 561, 461 A. 2d 184 (Law Div. 1983), *aff'd in part, rev'd in part,* 202 N.J. Super. 106 (App. Div. 1985), *cert. granted,* No. 24248 (N.J. Sept. 9, 1985); *Ayers v. Township of Jackson,* 1983 Env'tl Rep. (BNA) 1382 (N.J. Super. Ct. Law Div., Nov. 16, 1983).

[28] *Id.,* at 1382 (BNA).

[29] *Id.*

[30] *Id.*

[31] See *1,334 Former Love Canal Residents Receive $20 Million Under 1983 Settlement Agreement,* 1985 Env'tl Rep. (BNA) 1445 (Current Developments, Jan. 4, 1985).

[32] E.g., *Kenney v. Scientific Inc.,* 497 A. 2d 1310. (N.J. Super. Ct., Law Div. Apr. 3, 1985), ELR 15 20403; *Johnson v. E. I. duPont de Nemours & Co.;* No. 58339-80 (N.J. Super. Ct. Law Div.); *Anderson v. Cryovac, Inc.* No. 82-1672-S (D. Mass. filed May 11, 1982), ELR Pending Litigation 65758.

[33] *Barrage of Private Tort Claims Simmers Beneath Toxic Dumps,* Legal Times, Oct. 22, 1984, at 1, col. 2; Comment, *Hazardous Waste and the Common Law: Will New Jersey Clear the Way for Plaintiffs to Recover?,* 15 ELR 10321, 10323 at n. 16 (1985).

[34] See generally McGovern, *supra* note 5, at 9-16; Rosenberg, *supra* note 7, at 908-16; Phillips, *Practical Considerations in Undertaking Hazardous Waste Exposure Cases on Behalf of Multiple Plaintiffs,* in Hazardous Wastes, Superfund, and Toxic Substances Course of Study Materials 392, 400-01 (Env'tl L. Inst./ALI-ABA Comm. on Continuing Prof. Educ. 1984) (hereinafter cited as Phillips). For the management of asbestos cases see generally *Working Group Report, supra* note 9; *Asbestos Center,* 70 ABA J. 32 (July, 1984); *Asbestos Case Plan Submitted,* Legal Times, Oct. 24, 1983, at 1, col. 2; *Plaintiffs' Counsel Criticize Claim Facility,* Legal Times, Feb. 24, 1986, at 1, col. 2.

[35] Fed. R. Civ. P. 23. See also McGovern, *supra* note 5, at 9 and n. 18; Rosenberg, *supra* note 7, at 908-09 and nn. 223-226 *Court Upholds Class Action in Asbestos Case,* Legal Times, Mar. 3, 1986 at 5. Newberg; *Mass Tort Class Actions,* 22 Trial 53 (Feb. 1986).

[36] 28 U.S.C.A. § 1407. See McGovern at 11-12 and n. 19.

[37] *Id.,* at 12 and n. 20.

[38] Fed. R. Civ. P. 20(a).

[39] See 28 U.S.C.A. § 1404(a) (transfer and consolidation of cases originally filed in general districts); 28 U.S.C.A. § 1407 (transfer from coordinated pretrial to any federal district). See generally *Manual for Complex Litigation* (5th ed. 1982) (West). Unlike § 1404, transfer under § 1407 can be to any federal district court.

[40] 414 U.S. 291 (1973).

[41] Section 301(e) Report, *supra* note 4, in Part 2 at 343. This report recommends that Rule 23 status be granted to reflect the needs of tort victims (at 381). See also Rosenberg, *supra* note 7, at 908 n. 224.

[42] *Master Says Mass Torts Don't Belong in Courts,* Legal Times, Oct. 15, 1984, at 5, col. 3. See generally, Inst. for Health Policy Analysis, *Causation and Financial Compensation* (Georgetown U. Med. Center 1986) (hereinafter cited as *Financial Compensation*).

[43] Singer, *An Analysis of Common Law and Statutory Remedies for Hazardous Waste Injuries,* 12 Rutgers L.J. 117, 122 (1980) (hereinafter cited as Singer). See generally Last, *Tort Law Implications of Hazardous Waste Facilities,* 17 Nat. Resources Law 491 (1984) (reprinted in this volume are the papers from the Thirteenth Annual Airlie House Conference on the Environment) (hereinafter cited as Last).

[44] § 301(e) Report, *supra* note 4, Part 1 at 110-24.

[45] ELR § 301(e) Article, *supra* note 1, at 10099.

[46] *Black's Law Dictionary* 930 (5th rev. ed. 1979).

[47] Prosser, *supra* note 6, at 143.

[48] Comment, *Hazardous Waste Liability and Compensation: Old Solutions, New Solutions, No Solutions,* 14 Conn. L. Rev. 307, 322 (1982).

[49] 227 N.Y. 208, 210, 125 N.E. 93 (1919).

[50] Prosser, *supra* note 6, at 147-49, 157-68.

[51] *Id.,* at 145-49.

[52]   *Restatement (Second) of Torts* §§ 291-93 (1965). See *Compensation for Victims of Toxic Pollution —
        Assessing the Scientific Knowledge Base,* a report prepared for the Nat'l Sci. Found. (NSF) (1983)
        (hereinafter cited as *NSF Study*), reprinted in *Should Producers of Hazardous Waste be Legally Re-
        ponsible for Injuries Caused by the Waste?* H. R. Doc. No. 98-93, 98th Cong., 1st Sess. 223, 225(1983).
        (This document contains a compilation of readings and bibliographic references prepared by the Cong.
        Res. Serv. for use in the 1983-84 Intercollegiate Debate.)

[53]   See, e.g., Hurwitz, *Environmental Health: An Analysis of Available and Proposed Remedies for Vic-
        tims of Toxic Waste Contamination,* 7 Am. J.L. & Med. 61, 71 and n. 69 (1981) (hereinafter cited as
        Hurwitz).

[54]   493 F. 2d 1076, 1089 (5th Cir. 1973), *cert. denied,* 419 U.S. 869 (1974).

[55]   Baurer, *Love Canal: Common Law Approaches to a Modern Tragedy,* 11 Env'tl L. 133, 144-45 (1980)
        (hereinafter cited as Baurer).

[56]   *Ettyl* v. *Land and Loan Co.,* 122 N.J.L. 401, 404, 5 A. 2d 689, 691 (Sup. Ct. 1939).

[57]   267 F. 2d 425 (5th Cir. 1959). See generally Last, *supra* note 43, at 491-95 and accompanying notes for
        citations to recent cases.

[58]   See, e.g., *NSF Study, supra* note 52, at 224-26.

[59]   See, e.g., *Six Case Studies, supra* note 4, at 490; Strand, *The Inapplicability of Traditional Tort Anal-
        ysis to Environmental Risks: The Example of Toxic Waste Pollution Victim Compensation,* 35 Stan.
        L. Rev. 575, 583 (1983) (hereinafter cited as Strand).

[60]   *Id.,* at 582; S. Rep. No. 848, 96th Cong., 2d Sess. 2 (Sen. Comm. on Env'tl and Pub. Works 1980).
        Rosenberg, *supra* note 7 at 855 and nn. 26-27.

[61]   See, e.g., Black and Lilienfeld, *supra* note 9, at 735 n. 6; Rosenberg, *supra* note 7, at 855-58 and nn.
        28-41.

[62]   Statement by Dr. John Romani in S. Rep. No. 848, *supra* note 60, at 41.

[63]   See, e.g., *Six Case Studies, supra* note 4, at 488; *NSF Study, supra* note 52, at 223-24.

[64]   *Six Case Studies, supra* note 4, at 488. For an excellent discussion of legal vs. scientific causation, see
        Financial Compensation, supra note 42, at 11.

[65]   Black and Lilienfeld, *supra* note 9, at 735-38.

[66]   Hurwitz, *supra* note 53, at 71 n. 69.

[67]   See, e.g., Note, *Emotional Distress, supra* note 21, at 1215, n. 1.; Phillips, *supra* note 34, at 398.

[68]   Strand, *supra* note 59, at 583.

[69]   See Phillips, *supra* note 34, at 396-97 (citing *Feist* v. *Sears Roebuck & Co.,* 517 P. 2d 675 (Ore. 1973));
        Schwegel v. *Goldberg,* 209 Pa. Super. 280, 228 A. 2d 405 (1967); *Lindsay* v. *Appleby,* 91 Ill. App. 3d
        705, 414 N.E. 2d 885 (1980); *Figlar* v. *Gordon,* 133 Conn. 577, 53 A. 2d 645 (1947); *Starlings* v. *Suri
        Roundtop Corp.,* 493 F. Supp. 507 (M.D. Pa. 1980). Recovery for costs of medical surveillance (i.e.,
        diagnostic medical services reasonably required by persons wrongfully exposed to hazardous chemicals)
        is a theory of damages separate from enhanced risk (*id.,* at 397-98, citing *Schroeder* v. *Perkel,* 87 N.J.
        53, 432 A. 2d 834 (1981)); *Tyminski* v. *U.S.* 481 F. 2d 257 (3d Cir. 1973); *Wise* v. *Towse,* 366 S.W. 2d
        506 (Mo. 1963); *Whitney* v. *Akers,* 247 F. Supp. 763 (W.D. Okla. 1965); *City of Northglenn* v. *Chev-
        ron,* No. 81-Cr 44 (D. Colo. Apr. 21, 1982). For a detailed analysis of recent cases involving both
        traditional and evolving theories of damages in toxic tort cases see Rodburg, *Multi-Plaintiff Personal
        Injury Claims at Hazardous Waste Sites,* in Hazardous Wastes, Superfund, and Toxic Substances
        Course of Study Materials 271, 276-309 (Envtl. L. Inst./ALI-ABA Comm. on Continuing Prof. Educ.
        Oct. 1985).

[70]   E.g., *Ayers* v. *Township of Jackson, supra* note 27; *Dixon* v. *N.Y. Trap Rock Corp.,* 293 N.Y. 509,
        58 N.E. 2d 517 (1944); *McDonald* v. *Mianecki,* 79 N.J. 275 (1979); *Nitram Chemical Co.* v. *Parker,*
        200 S. 2d 220 (Fla. App. 1967) (where emotional injury awards are incidental to nuisance claim of
        groundwater contamination).

[71]   159 Eng. Rep. 737 (1865), *rev'd,* L.R. 1 Ex. 265 at 279-80 (1866), *aff'd,* L.R. 3 H.L. 330 (1868).

[72]   Singer, *supra* note 43, at 134-35; Prosser, *The Principle of Rylands v. Fletcher* in *Selected Topics on
        the Law of Torts* 158 (1953).

[73]   *Restatement (Second) of Torts* § 520 (1976); see also Last, *supra* note 43, at 499-502 and accompanying
        notes.

[74]   Baurer, *supra* note 55, at 139.

[75]   312 So. 2d 799,804 (Fla. App. 1975).

[76]   Prosser, *supra* note 6, at 512.

[77]   *Id.,* at 514.

[78]   Hurwitz, *supra* note 53, at 77.

[79]   *Six Case Studies, supra* note 4, at 108-09.

[80]   Gulick, *Superfund: Conscripting Industry Support for Environmental Cleanup,* 9 Ecology L.Q.
        524,533 (1981).

[81]   *NSF Study, supra* note 52, at 227.

[82]   See, e.g., Hurwitz, *supra* note 53, at 80.

83   *Restatement (Second) of Torts* § 402A. See also *Greenman* v. *Yuba Power Products, Inc.,* 59 Cal. 2d 57, 377 P. 2d 897, 27 Cal. Rptr. 697 (1963).

84   See, e.g., Hurwitz, *supra* note 53, at 76 n. 101; *Six Case Studies, supra* note 4, at 485.

85   Grad, *Remedies for Personal Injuries Resulting from Exposure to Hazardous Wastes* in *Environmental Law XII Course of Study Materials* 445, 463 (ELI/ALI-ABA Comm. on Continuing Prof. Educ. 1982) (hereinafter cited as Grad). See also Note, *Strict Liability for Generators, Transporters, and Disposers of Hazardous Wastes,* 64 Minn. L. Rev. 249 (1980).

86   *Six Case Studies, supra* note 4, at 485; *NSF Study, supra* note 52, at 225.

87   90 N.J. 191, 447 A. 2d 539 (1982).

87a 97 N.J. 429, 479 A.2d 374 (1984). See also *in re Asbestos Litigation,* Misc. No. 85-381 (D.N.J. 1986) (denial of state-of-the-art defense to manufacturers of asbestos products in strict liability actions does not violate Equal Protection Clause).

88   See, e.g., ELR § 301(e) Article, *supra* note 1, at 10099-100; Birnbaum and Wrubel, *The N.J. Supreme Court Breathes New Life into State-of-Art Defense,* Nat'l L.J., Sept. 17, 1984, at 22, col. 1.

89   Prosser, *supra* note 6, at 591-93, 595-99; see generally Last, *supra* note 43, at 496-97 and accompanying notes.

90   Grad, *supra* note 85, at 464.

91   See, e.g., Prosser, *supra* note 6, at 596-98; *Spar Industries Inc.* v. *Del E. Webb Dev. Co.,* 108 Ariz. 178, 494 P. 2d 700 (1972).

92   See Grad, *supra* note 85, at 464; *Six Case Studies, supra* note 4, at 481.

93   Prosser, *Nuisance Without Fault,* 20 Tex. L. Rev. 399, 411 (1942). See generally Bryson and MacBeth, *Public Nuisance, The Restatement (Second) of Torts, and Environmental Law,* 2 Ecology L.Q. 241 (1972); Last, *supra* note 43, at 497-99 and accompanying notes.

94   *State* v. *Schenectady Chemicals, Inc.,* 13 ELR 20550 (N.Y. Sup. Ct. 1983); *Copert Industries* v. *Consolidated Edison,* 41 N.Y. 2d 564, 362 N.E. 2d 968, 394 N.Y. S. 2d 169 (1977).

95   Grad, *supra* note 85, at 465; see also Prosser, *supra* note 6, at 604.

96   Grad, *supra* note 85, at 465-66.

97   *Nat'l. Sea Clammers Ass'n.* v. *City of New York,* 616 F. 2d 1222 (3d Cir. 1980), *rev'd sub nom. Middlesex County Sewerage Auth.* v. *Nat'l. Sea Clammers Ass'n.,* 453 U.S. 1 (1981).

98   See Grad, *supra* note 85, at 465-66.

99   See generally *Six Case Studies, supra* note 4, at 482; Prosser, *supra* note 6, at 594-96.

100 Hurwitz, *supra* note 53, at 81-82.

101 *Phillips* v. *Sun Oil Co.,* 307 N.Y. 328, 331, 121 N.E. 2d 249, 251 (1954).

102 See, e.g., Singer, *supra* note 43, at 126 n. 62.

103 Prosser, *supra* note 6, at 32.

104 *Six Case Studies, supra* note 4, at 483.

105 Hurwitz, *supra* note 53, at 80-81; Singer, *supra* note 43, at 126. See *Hall* v. *DeWeld Mica Corporation,* 244 N.C. 182, 93 S.E. 2d 56 (1956); *Martin* v. *Reynolds Metal Co.,* 221 Or. 86, 94 342 P. 2d 790, 794 (1959), cert. denied, 362 U.S. 918 (1960). See also Last, *supra* note 43, at 502-03 and accompanying notes.

106 Grad, *supra* note 85, at 462, citing *Eurrer* v. *Talent* Irrigation Dist., 358 Or. 494 466 P. 2d 605 (1971); *Rushing* v. Hooper-McDonald, 300 So. 2d 94 (Ala. 1974).

107 510 F. 2d 468 (4th Cir. 1975).

108 Prosser, *supra* note 6, at 292 states the principle of concerted action liability: "All those who, in pursuance of a common plan or design to commit a tortious act, actively take part in it, or further it by cooperation or request, or who lend aid or encouragement to the wrongdoer, or ratify and adopt his acts done for their benefit, are equally liable with him" (footnotes omitted). See *Restatement (Second) of Torts* § 876 (1977). See also Levine, *Gilding the Lilly: A DES Update,* 20 Trial 18, 22 n. 6 and accompanying text (Dec. 1984) (also suggesting a four-pronged approach toward identifying DES defendants).

109 The market share concept was first defined in *Sindell* v. *Abbott Laboratories,* 26 Cal. 3d 588, 607 P. 2d 924, 163 Cal. Rptr. 132 (Ct. App. 1979), *cert. denied* 449 U.S. 912 (1980): a DES plaintiff need only join those defendants comprising a "substantial percentage" of the market. For example, *Murphy* v. *E. R. Squibb and Sons, Inc.,* 710 P. 2d 247 (Cal. 1985) (no strict liability for pharmacists; 10% is not enough to meet the "substantial" market-share requirement of Sindell).

110 Illustrated by *Summers* v. *Tice,* 33 Cal. 80, 199 P. 2d 1 (1948) alternative liability applies where plaintiff can demonstrate that each defendant engaged in the tortious conduct causing injury. The burden then shifts to defendants to establish lack of fault; should defendants fail in their burden liability will be imposed on a joint and several basis.

111 See *infra* text accompanying note 113.

112 436 N.Y.S. 2d 625 (App. Div. 1981), *off'd.,* 436 N.E. 2d 182 (N.Y. 1982).

113 345 F. Supp. 353 (E.D.N.Y. 1972)

¹¹⁴   See Sindell, *supra* note 109; *Copeland* v. *The Celotex Corp.,* No. 81-977 (Fla. Dist. Ct. App. Mar. 6, 1984) (applying market share concept in asbestos case). See generally Trauberman, *supra* note 4, at 7 Harv. Env'tl L. Rev. 177, 230-34 and accompanying notes.

¹¹⁵   418 Mich. 311, 343 N.W. 2d 164 (Mich. 1984), *cert. denied,* No. 83-2093, 53 U.S.L.W. 3227 (Oct. 1, 1984). See ELR § 301(e) Article *supra* note 1, at 10099. Levine, *supra* note 108, at 19, explains that under *Abel* plaintiff still must clear three evidentiary hurdles: (1) all defendants acted tortiously; (2) plaintiff was harmed by the conduct of one defendant (requiring all possible defendants to be joined); and (3) through no fault of plaintiff the actual manufacturer cannot be identified.

¹¹⁶   116 Wis. 2d 166, 342 N.W. 2d 37 (Wis. 1984), *cert. denied,* No. 83-1932, 53 U.S.L.W. 3226 (Oct. 1, 1984). Under this "risk contribution" theory: plaintiff need only sue one defendant; that defendant may proceed against other manufacturers for contribution; and liability will be shared among defendants who are unable to prove that they could not have made the DES in question.

¹¹⁷   See § 301(e) Report, *supra* note 4, Part 1 at 43-45, 133 n. 4, 255-56, 261 and Part 2 at 13-78. The § 301(e) Report notes that "[a]t least thirty-nine states have adopted a "discovery rule" in some form (footnote omitted)." *Id.,* at 43. See Note, *The Fairness and Constitutionality of Statutes of Limitations in Toxic Tort Suits,* 96 Harv. L. Rev. 1683 (1983) which concludes: "Toxic tort statutes of limitations therefore violate fundamental principles of justice implicit in the individual's relation to society" (*id.,* at 1702), and urges courts and legislatures to protect the rights of injured plaintiffs. For a comprehensive outline of state statutes of limitations, see *In re Agent Orange Product Liability Litigation,* 597 F. Supp. 740, 880-98 (EDNY 1984).

¹¹⁸   Birnbaum, *Statutes of Limitations in Environmental Suits:* The Discovery Rule Approach, 16 Trial 38, 38 (1980). See, e.g., *Statute of Limitations Expanded in DES Cases,* (*Kensinger* v. *Abbott Labs,* 217 Cal. Rptr. 313 (Cal. Ct. App. 1985)), Legal Times, Sept. 9, 1985 at 7.

¹¹⁹   *Abram* v. *Occidental Chemical Corp.,* Env'tl Rep. (BNA), July 8, 1983, at 385 (Niagara County, N.Y. June 6, 1983) (memorandum decision).

¹²⁰   See, e.g., Grad, *supra* note 85, at 457-58.

¹²¹   15 U.S.C.A. §§ 2601-29 (1976). See Kahan, *Reporting Substantial Risks Under Section 8(e) of the Toxic Substances Control Act,* 19 B.C.L. Rev. 859 (1978).

¹²²   15 USCA at § 2604(b).

¹²³   519 F. 2d 31 (5th Cir. 1975), *cert. denied,* 425 U.S. 971 (1976).

¹²⁴   FIFRA, 7 USCA §§ 135-136y (1970, as amended by the Federal Pesticide Act of 1978); Food, Drug and Cosmetic Act, 21 USCA §§ 301-392 (1962); OSHA, 29 USCA §§ 651-678 (1970); FWPCA, 33 USCA §§ 1251-1376 (1972, as amended by the Clean Water Act, 1977); FSDWA, 42 USCA §§ 300f-300j-10 (1977); Atomic Energy Act, 42 USCA §§ 2011-2282 (1946, 1954); RCRA, 42 USCA §§ 6901-6987 (1976, as amended by the Hazardous and Solid Waste Amendments of 1984); CAA, 42 USCA §§ 7401-7642 (1970); CERCLA, 42 USCA §§ 9601-9657 (1980).

¹²⁵   *Usery* v. *Turner Elkhorn Mining Co.,* 428 U.S. 1 (1975); The Black Lung Benefits Act of 1972, 30 U.S.C.A. § 801, 901-41 (1976 and Supp. III 1979).

¹²⁶   633 F. 2d 1212 (6th Cir. 1980), modified on rehearing, 657 F. 2d 814, *cert. denied,* 454 U.S. 1109 (1981).

¹²⁷   *Eagle-Picher Industries, Inc.* v. *Liberty Mutual Ins. Co.,* 682 F. 2d 12 (1st Cir. 1982), *cert. denied,* 103 S. Ct. 1279 (1983).

¹²⁸   *Keene Corp.* v. *Ins. Co. of North America,* 667 F. 2d 1034 (D.C. Cir. 1981), *cert. denied,* 455 U.S. 1007 (1982).

¹²⁹   *American Home Products* v. *Liberty Mutual Ins.,* 565 F. Supp. 1485 (S.D. N.Y. 1983).

¹³⁰   No. 82-0669 (D.D.C. Apr. 12, 1984).

Chapter 8

THE LEGAL DEVELOPMENT OF REMEDIES OF HAZARDOUS WASTE VICTIMS — STATUTORY REMEDIES*

Kenneth G. Bartlett

TABLE OF CONTENTS

* All footnotes appear at end of chapter.

In turning to various legislative and administrative bodies for available remedies, there is not a full compliment of statutory support for obtaining redress for hazardous waste and toxic injuries. The following outlines the various types of legislative proposals which continue to emerge in this area.

I. TYPES OF STATE STATUTES AFFECTING PRIVATE LITIGATION

In establishing the regulatory scheme for state control of potential polluters, state legislatures may either affirm, repeal, or ignore any common law cause of action for damages already recognized in that state.

The langugage of the statutes will provide subsequent judicial interpretations with evidence of legislative intent which will then shape or deny private redress. Thirteen states have provisions in their environmental statutes which expressly state that the policy of those acts is not to preempt or change the common law and that any prior right to private action should continue in coexistence with state statute.[1]

These statutes basically provide that any redress by the state provided by these environmental statutes is in addition to any remedy existing at common law.

Another type of statute is one that neither limits preexisting causes of action, nor provides new rights to private litigants; it *expressly* has no effect on the common law at all. Typical is the Wyoming Environmental Quality Act.[2] It is intended to be purely regulatory in its nature and effect and is not to impact on private suits.[3]

Lastly, state statutes may make no reference whatsoever to private litigation or redress from injuries due to hazardous waste. Under this latter type of statute, courts would be the most free to liberally construe the act and to apply the legislative intent to the facts of the particular case before it.

Generally, in order to imply a cause of action into a state law, courts will examine the underlying legislative intent, or if none is discernible, the apparent purpose of the statute. It is largely the same policy considerations in implying a cause of action that occurs under a federal statute that are applicable here.

The usual methodology for the plaintiff is to allege that he is a member of the class for whose benefit the statute was enacted and that the harm suffered is the type that the statute was meant to prevent. Of course, the theory of implying a private cause of action into a state regulatory statute presupposes that no mention is made of either providing or denying a private plaintiff the right to bring his own legal action.

II. STATUTES ASSISTING THE PRIVATE LITIGANT

Once the injured party is in court, either under a preexisting common law theory or a new statutory right of action (or, in the very least, was not deprived of his right to sue privately by his state's environmental statute), he may be provided with other statutory assistance in trying his case.

A central theme developed in earlier chapters is the stringent proof required and the high costs of obtaining such proof on issues of causation. A state statute may, in fact, ease this burden in a variety of ways. It is generally recognized that the most significant obstacle in a toxic pollutant case is proving causation. The prevailing rule in most jurisdictions is that in addition to identifying the defendant, the plaintiff must show by the preponderance of the evidence that the conduct of that particular defendant was a "substantial factor" leading to the injury. It is self evident that the burden becomes increasingly difficult to sustain against any one defendant when the number of defendants increases. The nature of proof in this type of suit pales when compared with proof typically offered in a traditional tort case. Virtually every hazardous waste/toxic pollutant case requires expert testimony on the issue of causation. Since many environmental cases turn on technical issues, expert testimony often dominates the trial. Proof will be compounded in difficulty when there are multiple defendants or when there is a long latency period, or both.

While the state of medical science continues to advance, proof of causation through model studies is complex, costly, and time-consuming. This difficulty is very often heightened by experts themselves disagreeing, creating an overlay of credibility to the factual and legal conclusions reached.

A. Negligence Per Se

Negligence per se is a doctrine which bases a finding of breach of a duty upon the violation of a statutory standard of conduct. Use of a violation of standards of conduct to show negligence is not new, nor limited in application to a waste litigation context. It has been part of the common law of torts for many years. As Dean Prosser notes:

> Once the statute is determined to be applicable — which is to say, once it is interpreted as designed to protect the class of persons in which the plaintiff is included, against the risk of harm which has in fact occurred as a result of the violation — the great majority of the courts hold that an unexcused violation is conclusive on the issue of negligence and that the court must so direct the jury.[4]

However, the plaintiff still must present evidence on the issue of proximate cause; that is, that a causal connection exists between the particular statutory violation and the plaintiff's injuries. Additionally, the plaintiff must rebut any claim made by the defendant that the plaintiff either was contributorily negligent or contributory negligence or assumed the risk. In other words, the doctrine of negligence per se only establishes one aspect of the plaintiff's case; by no means does it provide him with a judgment in his favor.

Some states which have not totally accepted a negligence per se doctrine have reached almost the same result by holding that statutory violations establish a presumption of negligence. The defendant is free to rebut this presumption. Since the laws differ from state to state, one has to examine each state's civil procedure laws to determine the type of jury instructions given in such a case.

Lastly, a small minority of states hold that evidence of statutory violation is not conclusive nor presumptive, and can be used only inferentially to establish breach of the duty owed the plaintiff. The jury, in this latter case, may weigh and apply it to the issues as it sees fit.

In federal actions, the statute based standard of care may also be of assistance in proving negligence. In *Orthopedic Equipment Company* v. *Eutsler,*[5] the Fourth Circuit Court of Appeals found a federal food, drug, and cosmetic act violation to be negligence per se under Virginia law.

It should be noted that the converse of the above doctrine is not necessarily valid. The courts have been uniform in holding that compliance with a statute's standard of care is not a complete defense in a negligence case. "The statutory standard is no more than a minimum, and it does not necessarily preclude a finding that the actor was negligent in failing to take additional precautions."[7]

In other words, courts are of the view that a statutory standard of care establishes only a minimum standard required. The existence of "special circumstances" would render the statutory standard inapplicable as the proper behavior given the existence of these additional circumstances.[8] One could expect, however, to meet this type of defense in hazardous waste cases and should consider and prepare to prove the existence of special or extenuating circumstances.

B. Strict Liability[9]

State statutes may also provide the hazardous waste litigant with doctrinal support by imposing the theory of strict liability upon producers, transporters, or storers of hazardous waste. Strict liability is a doctrine which imposes liability upon a responsible person, even though he is not responsible of any wrongdoing or deviation from some prescribed standard of care. Basically, it is a theory of social policy that intends to place the burden of financial loss upon the person engaged in some particular type of activity. In essence, it is the price one must pay for doing business of the sort that the defendant is engaged in.

As was mentioned, strict liability is imposed upon a defendant without the usual stigma of "wrongdoing" and without the normal showing of fault. As Prosser notes:

> The problem is dealt with as one of allocating a more or less inevitable loss to be charged against a complex and dangerous civilization, and liability is imposed upon the party best able to shoulder it. The defendant is held liable merely because, as a matter of social adjustment, the conclusion is that the responsibility should be his.[10]

This doctrine was first applied in the so-called blasting cases where injury and damage occurred despite the highest degree of care taken. At the time of this writing, four states have statutes expressly applying strict liability theories to injuries resulting from toxic wastes. Under Alaska law,[11] injured parties may recover for injuries which include personal or property losses or losses due to interruption of income earning potential resulting from toxic spillage. The defendant is held strictly liable in absence of relatively strict special circumstances.[12]

North Carolina, Maine, and Rhode Island also have enacted laws directing parties to be strictly liable for injuries resulting from hazardous or toxic substances which either spill into state waters or are improperly disposed of.[13] These latter states do not provide redress or apply the strict liability doctrine to as wide a range of injuries as does Alaska. Each state should be specifically checked for applicability.

C. Nuisance Laws

Approximately one half of the states have enacted nuisance statutes which, depending on their language, may be of assistance to the private litigant injured by the exposure of toxics. These state laws are in addition to any rights existing under nuisance common law which the state follows. Nuisance, divided into private or public, has been defined as:

> The essence of a private nuisance is an interference with the use and enjoyment of land. The ownership or rightful possession of land necessarily involves the right not only to the unimpaired condition of the property itself, but also to some reasonable comfort and convenience in its occupation.[14]

The prime limitation with private nuisance, as noted in the previous chapter, is that it requires the complaint to show that he has experienced a particularized or unique injury. Public nuisance, on the other hand, is an interference of the use and enjoyment of land affecting the community at large and usually is remedied by an action brought by the state. Virtually every state has provisions for enforcement and abatement of a public nuisance.

The Restatement has adopted the position that "in order to recover damages in an individual action for a public nuisance, one must have suffered harm of a kind different from that suffered by other members of the public exercising the right common to the general public that was the subject of the interference."[15] Of the 24 states that have enacted nuisance statutes, some track the restatement view that the plaintiff must show some sort of special injury before suing under the statute. These statutes give a private right to recover damages. This expands the remedy at common law, which was limited to abatement.

The successful plaintiff may seek equitable or legal remedies against the defendant, i.e., abatement of the nuisance and damages for his diminished use and enjoyment of his property.[16]

D. The Parties

One way in which a state may ease the burden on plaintiffs in hazardous waste litigation is to enact statutes which deal with an apportionment of liability among two or more possible defendants. Where the waste disposal area or facility has changed hands or control and the wastes have been in place for an extended period of time, the plaintiff, in the general case, has a fairly stiff burden of proportioning liability among the various owners and identifying the capable party.

Two states have eased this burden by providing that all persons who have or had control over the hazardous substances at a given point in time are to be jointly and severally liable for any injuries. Both Massachusetts and North Carolina have responded to the plaintiff's plight in this matter.[17]

E. Rebuttable Presumptions of Causation

One problem usually associated with hazardous waste litigation is the generic nature of the harmful substances involved. If a water table or stream becomes infected, the injured plaintiff must point to the responsible party with reasonable certainty.

A few more progressive states have eliminated this burden. Pennsylvania and Massachusetts have created rebuttal presumptions establishing causation in their environmental statutes.[18] Any spill or leakage is presumed to cause the injuries documented by parties near the areas in question.

Minnesota's Superfund law was enacted into law in April, 1983. The state's superfund law establishes a statutory strict inability cause of action for any past or future economic loss, death, injury, or disease caused by exposure to hazardous waste. This act contains a statutory rebuttable presumption establishing causation. Its retroactivity back to 1973 (and in some cases back to 1960) has caused considerable controversy, prompting some proponents to lobby the establishment of a state insurance pool.

In February 1984, a family in West Chester, Pa. was awarded $788,867.00 against an unlicensed waste hauler. It was alleged that the family's water supply had been polluted by toxic wastes dumped on an adjacent farm. This action in *Dietz* v. *Gray Brothers* had been brought under Pennsylvania's solid waste management act provision.[19]

This is a major step in that by shifting the burden of proof, plaintiffs are not inhibited by the costly and technical elements of showing causation in cases of hazardous waste injuries.

F. Federal Victim's Compensation

As noted earlier, primary drafts of CERCLA contained provisions where private individuals would be compensated for injuries caused by toxic pollutants.[20] This provision was deleted in the passage of this legislation in December 1980; it remains an important milestone in the social development of an alternative compensation system.

The release in September 1982 of the § 301(e) report has intensified debate on the efficacy of national legislation establishing a uniform federal statutory remedy to private individuals.[21] This September 1982 report has concluded that "available remedies are inadequate in view of the substantial number of claims that may arise, and the factual and legal complexities that will be invoked in their litigation.[22]

Senate bill 2892 (introduced July 31, 1984 to reauthorize CERCLA) contained language allowing for a limited federal cause of action for personal injury caused by toxic waste exposure. This provision contemplated recoveries for objectively determined expenses such as medical costs and loss of earnings, but not pain and suffering. This provision, which was deleted by an amendment to S. 2892, was opposed by both the U.S. Justice and Commerce Departments.

The Superfund report has made several recommendations concerning the use of this procedural device to ease the proof problem. The report has made ten recommendations concerning the establishment of a two-tier compensation system. It would be created through federal law and managed by the state. Compensation under this first tier would be analogous to compensation presently awarded under various state workers' compensation law[23] (e.g., medical expense and loss of earnings, but not pain and suffering). It would allow limited compensation through an administrative process without requiring the showing of fault to those injured by the disposal, transportation, and management of hazardous waste.[24] Monies paid out under this first tier would be financed by taxes on the production of toxic chemicals and crude oil.

G. Damages

Lastly, a state may aid an injured person and encourage waste-related suits by providing additional recovery to successful plaintiffs. New Hampshire, for example, provides treble damages if the defendant has been found in violation of that state's environmental protection statutes and violation has resulted in the plaintiff's injuries.[25]

III. STATE FINANCIAL RESPONSIBILITY LAWS

As important as it is to be provided with a cause of action, it is equally important to have a responsible defendant or a "deep pocket" from which to recover damages.

States have responded to this aspect of hazardous waste litigation basically in two ways: by providing a pool or fund from which a plaintiff may draw and by requiring hazardous waste operators to maintain insurance or provide some type of performance bond.

A. Fund Statute

Sixteen states have set up funds for emergency clean-up of hazardous waste spills.[26] These funds are not available to private individuals as a source of recovery for personal damages or injuries except to the extent that the polluted areas may be restored. Florida and New Jersey have come a step beyond in that they allow injured parties to make claims against the funds for personal injuries resulting from release of hazardous waste.[27]

In the case of Florida and New Jersey, the fund is a source to which injured individuals may look for recovery. It is administered by the state environmental protection agency which contains its own standards of proof and causation. After recovery, the state is subrogated to the injured party's claims against the wrongdoer. In essence, this type of statute transfers the time and expense of litigation from the private litigant to the state.

B. Operator Responsibility Law

Many states under Interim RCRA Authorization provide that an owner of a waste disposal site must maintain liability insurance or provide a bond sufficient to cover potential liability in the case of a spill or accidental discharge. The statutes range from expressly requiring for the financial ability of operators to pay private plaintiffs for resulting injuries[28] to only insuring the operation itself remains fiscally sound.[29] Between these two extremes, five states require the owner or operator to provide for liability, though these state statutes do not define the extent of liability, or the types of injuries protected by that law.[30]

IV. FEDERAL STATUTES — IMPLYING A PRIVATE CAUSE OF ACTION

As discussed in the previous chapter, there are numerous federal acts which prescribe, define, or require specific conduct with respect to hazardous wastes. Some provide relief for private litigants and others are purely regulatory. In any case, where injuries have occurred, a plaintiff should investigate the possibility of recovery under a given statute by attempting to imply into it a private cause of action.

The most illustrative case to date is *Cort v. Ash*[31] where the Supreme Court was called upon to decide whether a private cause of action existed under the Occupational Safety and Health Act (OSHA) of 1970. The court listed four factors to consider in determining whether a private cause of action can be implied in a statute not expressly providing for one. These four factors are

1. Is the plaintiff one of the class for whose especial benefit the statute was enacted?
2. Is there any indication of legislative intent, explicit or implicit, either to create such a remedy or to deny one?
3. Is it consistent with the underlying purpose or legislative scheme to imply such a remedy for the plaintiff?
4. Is the cause of action one traditionally relegated to state control and state remedies?

Application of these factors would result in finding of a private, vis-a-vis and administrative, remedy. As was discussed, *supra,* in the state statute section, a given federal statute might also be used to establish a standard of strict liability. Violation would, of course, be tantamount to a *prima facie* case of negligence. Each federal state, of course, must be analyzed within the above framework.

V. FEDERAL STATUTES PROVIDING FINANCIAL RECOVERY

A. Social Security Disability Insurance (42 U.S.C. 401 *et. seq.*)

Medicare is basically a national health insurance program for the aged and disabled. Basically, it covers individuals who have reached the age of 65 and who are entitled to receive social security retirement benefits or to individuals who are under the age of 65 and are entitled to receive disability benefits.

Medicaid, on the other hand, provides medical assistance through the state to individuals who are considered to be economically disadvantaged, i.e., families who, because of a special need, are not able to meet the costs of medical services. Funds are made available to states which have established a program attempting to reach the given disadvantaged group.

B. Black Lung Benefits (30 U.S.C. 901 *et. seq.*)

The Black Lung Act provides benefits to miners who are totally disabled by respiratory ailments and to the surviving dependants of miners who are totally disabled at the time of death. The benefits provided are available through a combination of federal funds, state workers' compensation laws, and payments which were made directly to mine operators. In addition to the benefits available, the Act also provides numerous rebuttable presumptions of the type as those discussed *supra.* Additionally, the Act also provides for financial responsibility by mine operators. A trust fund is established and administered under the auspices of the Secretary of Labor.

C. Federal Employees Liability Act (45 U.S.C. 54 *et. seq.*)

Federal Employees Liability Act basically operates to eliminate preexisting common law barriers to recovery by government employees for injuries due to negligence on the part of the U.S. Government.

D. Longshoreman's and Harbor Worker's Compensation Act

The Longshoreman's and Harbor Worker's Compensation Act provides compensation for death or disability occurring on the navigable waters of the U.S. If the death or disability arises out of injury incurred while the individuals were engaged in merit time employment, compensation includes payment for total disability, permanent or temporary, and is set by a schedule linked to the type of disability. Death benefits are also available to surviving spouses and dependants.

E. Compensation for Work Injuries — Government Employees (5 U.S.C. 8101 *et. seq.*)

The federal government also provides compensation for death or disability of government employees resulting from personal injury sustained all in the performance of government employment. This section of the Code does not apply to acts of willful misconduct or when the injuries are self-inflicted or the result of inebriation.

F. Veteran's Benefits (38 U.S.C. 101 *et. seq.*)

Veteran's Benefits provide compensation for death or disability or for any disease incurred in the line of military duty with the exception of self-inflicted injuries. To recover, a former serviceman must have incurred the injury in the course of the performance of his military duties, must have been honorably discharged from the armed forces, and the injury must become apparent within 1 year after discharge.

G. The War Risk Hazard Compensation Act (42 U.S.C. 1701)

These sections of the U.S. Code provide for compensation of employees or government contractors or "nonappropriated fund instrumentalities" such as army post exchanges, officers clubs, etc., for injuries resulting from war risk hazards. A war risk is considered to be any hazard arising during a war in which the U.S. is engaged, and which results from the discharge of any type of missile.

H. Federal Torts Claims Act (28 U.S.C. 1346(b))

Basically, the Federal Torts Claims Act waives the sovereign immunity generally attributed to the U.S. Government and bestows upon the Federal Courts exclusive jurisdiction to hear claims against the U.S. for money damages arising from tortious conduct of government employees when acting within the scope of their employment.

The Act provides that the U.S. be liable in the same manner and extent as any private individual in similar circumstances.

Questions which arise as to the liability of the government for the acts or omissions of contractors or subcontractors are referred to the laws of the state in which the tort occurs. Analysis which is similar to agency analysis is usually applied in these situations.

I. Price Anderson Act (42 U.S.C. 2201 2222)

The Price Anderson Act is basically an Act which requires that funds are available to compensate injured parties in the event of a nuclear accident. Potential liability of a single nuclear plant licensee is limited to $560 million for claims arising from a single extraordinary nuclear occurrence. Under the Act, plaintiffs would recover under a strict liability theory and the defenses of contributory negligence, government immunity, and statute of limitation are waived. Certain sections of the Price Anderson Act are designed to spread the potential costs of a nuclear accident among all operating members of the industry.

VI. CONCLUSION

Though this review of several of the Federal Acts available to a plaintiff for recovery of certain specific injuries is not meant to be exhaustive, it does illustrate that the sources of funds are many and varied. Additionally, there are numerous other acts providing funds or revenue sources from which an injured party may draw, and it is recommended that specific injuries and unusual tortious acts be investigated further.

FOOTNOTES

1. Hawaii Rev. Stat. § 342-16 (1976); Idaho Code § 39-108(8) (1977); Ill. Ann. Stat. ch. 111-½ § 1022.3 (Smith-Hurd) (Supp. 1981-82); Ky. Rev. Stat. § 224.995(3) (1977); La. Rev. Stat. Ann. § 30:1074(4) (West) (Supp. 1982); Me. Rev. Stat. Ann. tit. 38, § 1306-c(5) (Supp. 1981-82); Md. Nat. Res. Code Ann. § 8-1403 (1974); Mo. Ann. Stat. § 260.425(7) (Supp. 1980); Nev. Rev. Stat. § 445.321 (1979); N.M. Stat. Ann. § 74-6-13 (1981); N.D. Cent. Code § 32-40-04 (1976); Pa. Stat. Ann. tit. 35, § 6018.607 (Purdon) (Supp. 1981); and Tex. Rev. Civ. Stat. Ann. art. 4477-7 § 10 (Vernon) (Supp. 1982).

2. Wyo. Stat. § 35-11-901 (1982 Supp.).

3. It has not been decided however, even with so clear an expression of legislative intent, whether courts will imply a per se theory of negligence when it is violated or not.

4. Prosser, W. L., *Handbook of the Law of Torts,* West Publishing Co., St. Paul, Minnesota, 1971, Section 36.

5. 276 F. 2d 455 (4th Cir. 1960).

6. 21 U.S.C.A. § 301 *et. seq.*

7. Prosser, *Law of Torts,* § 36.

8. See *Restatement of Torts,* Section 288C and Comments.

9. For discussion of the application of strict liability of the toxic tort and hazardous waste situation, see, e.g., Baurer, T., Love Canal: Common Law Approaches to a Modern Tragedy, 11 Env'tl L. 133, 138-142 (1980); Ginsberg, W. H. and Weiss, L., Common Law Liability for Toxic Torts: A Phantom Remedy, 9 Hofstra L. J. 859 (1981); Milhollin, G., Long Term Liability for Environmental Harm, 41 Pittsburgh L. Rev. 1, 6-8 (1979); Pfennigstorf, W., Environment, Dangers and Compensation, 1979 A.B.F. Res. J. 349, 373-375; Note, The *Rylands v. Fletcher* Doctrine in America: Abnormally Dangerous, Ultrahazardous, or Absolute Nuisance? Arizona S.L.J. 99 (1978); Note, Pursuing a Cause of Action in Hazardous Waste Pollution Cases, 29 Buffalo L. Rev. 533, 1980.

10. Prosser, *Law of Torts,* § 75 (1971), 4th ed.

11. Alaska Stat. 46.03.824 (1977).

12. E.g., War, Acts of God, U.S. Acts of Negligence by the State or Third Party Intervention, Alaska Stat. 46.03.822 (1977).

13. See N.C. Gen. Stat. § 143-215.77 (1978), *et seq.,* Maine Rev. Stat. Sec. 38, Section 1306-1 (5) (1981 Supp.), R.I. Gen. Laws Section 23-19.1-22 (1980 Supp.).

14. Prosser, *Law of Torts,* § 89 (1971), p. 591.

15. Hazardous Wastes: Preserving the Nuisance Remedy, 33 Stan. L. Rev. 675 (1981).

16. See generally, French, B., Private Nuisance Approach to Hazardous Waste Disposal Sites, 7 Ohio N.U.L. Rev. 86 (1980).

17. Mass. Ann. Laws, ch. 21, Section 27 (14) (1980); N.C. Gen. Stat. Section 143-215.94 (1979 Supp.).

18. P.A. Stat. tit. 35, Section 6018 (1982 Supp.); Mass. Ann. Laws, ch. 21 § 10 (1980).

19. *Dietz* v. *Gray Brothers,* Chester County, Ct. Common Pls., No. 16 (1980).

20. On November 24, 1980, the Senate version of the Superfund (S. 1980) was indefinitely postponed in lieu of the House bill (H.R. 7020) Comprehensive Environmental Responses, Compensation and Liability Act of 1980, Pub. L. No. 96-510 (1980). "The . . . more ambitious $4.1 billion measure was cut back about 75 percent to appease the Republican opponents who had threatened to filibuster it." (1980) Cong.

Q. (CCH) 34-35. For legislative history, see H. Rep. No. 96-1016, 96th Cong., 2nd Sess. Reprinted in (1981) U.S. Code Cong. & Ad. News 10249.

[21] See § 301(e), Report of the Comprehensive Environmental Response, Compensation and Liability Act of 1980 (serial No. 97-12).

[22] § 301(e) Report at

[23] "The Tier One compensation system is a risk-sharing system similar in this approach to such older, employment-related, insurance as workers' compensation." 301(e) Report at 180.

[24] 301(e) Report at 192.

[25] N.H. Rev. Stat. Ann. Section 147: 59(i) (1981 Supp.).

[26] Alabama, Arizona, Colorado, Connecticut, Florida, Georgia, Illinois, Louisiana, Maryland, Michigan, New Hampshire, New Mexico, North Carolina, Pennsylvania, Tennessee, and Wisconsin.

[27] Fla. Stat. Sec. 403.725(1) (1981 Supp.); N.H. Rev. Stat. Sec. 58: 23.11g(A) (1982).

[28] Fla. Stat. 403.724(3) (1981 Supp.), Okla. Stat. tit. 63, Sec. 1-2008(B) (1981-1982).

[29] Alaska Stat. 46.03-830 (1981 Supp.); Ga. Code Ann. 43-2909(1) (1981 Supp.); Ky. Rev. Stat. annotated Sec. 224.844 (1981 Supp.); Md. Nat. Res. Code Ann. 8-1413.2 (1981 Supp.); Nev. Rev. Stat. Sec. 445.294 (1979); Utah Code Ann. Sec. 26-14-8 (4) (4) (C) (1981 Supp.).

[30] Miss. Code Ann. § 17-17-27 (1981 Supp.); Neb. Rev. Stat. § 81-1521.04 (1980 Supp.); N.H. Rev. Stat. Ann. § 147-A: 5 (1981 Supp.); P.A. Stat. Ann. tit. 35. § 2018.506 (1981 Supp.); and Tenn. Code Ann. 53-6308 (1981 Supp.).

[31] 422 U.S. 66, 95 S. Ct. 2080.

Chapter 9

ECONOMIC ISSUES AND ASPECTS OF HAZARDOUS WASTE MANAGEMENT

Gilah Langner, Steve Bailey, David Bruce, and Robin Rodensky

TABLE OF CONTENTS

I. INTRODUCTION

Recent years have seen rapid changes in the field of hazardous substance and hazardous waste management in a variety of industries. The primary impetus for these changes has come from federal legislation and associated regulations under the Resource Conservation and Recovery Act and the Superfund law administered by the U.S. Environmental Protection Agency (EPA). To a significant degree as well, states have begun to assume a larger role in this field, implementing authorized RCRA programs and passing Superfund legislation of their own. This introduction provides an overview of the RCRA and Superfund laws and regulations affecting industry and reviews the major sources of current and potential costs resulting from those laws and regulations.

The Resource Conservation and Recovery Act (RCRA) was signed into law on October 21, 1976 and subsequently amended most recently on November 8, 1984. Built on the foundation of the Solid Waste Disposal Act of 1970, RCRA represents a far-reaching answer to the problem of solid waste management. Along with other environmental legislation enacted at the time, such as the Toxic Substances Control Act and the Safe Drinking Water Act, RCRA emphasizes prevention rather than cleanup; recognizes the complexity of engineering and management solutions to environmental problems; and takes a multi-media approach towards environmental quality.

RCRA embodies these goals in two key provisions. The Act creates: (1) a new permit program covering every hazardous waste disposal site, and (2) a "cradle-to-grave" manifest system to track every load of hazardous waste material from generation at a plant to ultimate disposal.

Because of the extensive controls instituted for the handling of hazardous wastes, the definition of the term "hazardous" under Subtitle C of RCRA was both complex and controversial. RCRA defines hazardous wastes as those solid wastes which, by their quantity, concentration, or physical, chemical, or infectious characteristics, may contribute to mortality or illness, or pose a hazard to human health or the environment. Subtitle C of RCRA mandates EPA to identify the characteristics of hazardous waste and to list particular hazardous wastes which will be subject to regulation. EPA has issued regulations under Section 3001 that contain criteria under which a waste is to be considered hazardous and that also specifically list particular hazardous wastes in the following three categories: hazardous wastes from nonspecific sources, hazardous waste from specific sources, and discarded commercial chemical products.

As will be discussed further below, several other aspects of the RCRA-mandated hazardous waste management system will affect generators, transporters, and owners and operators of hazardous waste disposal facilities, including record-keeping, reporting, and financial responsibility requirements.

The Comprehensive Environmental Response, Compensation and Liability Act of 1980 (CERCLA), more commonly known as the *Superfund law,* establishes a trust fund of $1.6 billion to finance cleanups of abandoned hazardous waste sites and ongoing emergency releases. The money for the fund comes from general revenues and taxes on crude oil and certain chemical feedstocks. CERCLA, Section 104 establishes broad federal authority to respond to releases or threats of releases of hazardous substances, pollutants, or contaminants from vessels and facilities. Hazardous substances are defined to include all substances designated as such under other federal statutes, while pollutants or contaminants cover a broad range of other possibly harmful substances. The blueprint for conducting government response actions is contained in the National Contingency Plan, which was most recently revised and published by EPA on November 20, 1985.

Other relevant sections of the law are briefly mentioned here. Sections 102 and 103 of CERCLA provide for the designation of additional substances as hazardous, the assignment of reportable quantities to hazardous substances, and the required reporting of all releases into the environment of hazardous substances in an amount equal to or greater than the reportable quantity. The notification requirements contained in Section 103 are applicable both to episodic and to continuous releases of hazardous substances. CERCLA also establishes authority to take enforcement actions (Section 106) to compel cleanup of hazardous substance releases by responsible parties or to recover costs of response from such parties. The liability provisions contained in CERCLA are set forth in Section 107 and financial responsibility provisions are included in Section 108.

The Superfund tax on petroleum and certain chemicals, which is the major source of funding for the trust fund, expired at the end of September 1985. The substantive provisions of the statute, however, remain in effect.

Congress, as of early 1986, is considering Superfund reauthorization legislation, which would increase the size of the trust fund and alter some of the substantive provisions of the statute.

RCRA and Superfund represent an interesting contrast in Congress' attempt to control problems brought about by hazardous chemicals and wastes. The two laws differ both in the aspects of the problem that they deal with and in the solutions that they offer. RCRA deals with the on-going generation, transportation, treatment, storage, and disposal of hazardous waste; Superfund deals with the past legacy and inevitable emergencies associated with hazardous susbtances. Thus, for the most part, Superfund is designed to clean up problems already occurring by means of a trust fund. RCRA is designed to establish a system to prevent future need for Superfund by properly managing hazardous waste at each point in the chemical cycle.

The two laws differ more fundamentally in the solutions and incentives they provide. CERCLA operates primarily through establishment of liability for response costs and natural resource damages; RCRA operates primarily through the regulatory process. Both approaches are aimed at producing changes in regulated parties, in the form of an increase in the standard of care exercised by industry in handling hazardous substances. An increase in the standard of care can either be directly required, through the RCRA permitting process, for example, or indirectly induced through the establishment of stringent liability provisions, as under CERCLA. By increasing the amount of damages that firms would be liable for and the probability that a release would result in successful litigation by injured parties, such liability provisions create incentives for firms to prevent the release of harmful substances. However, it has not yet been investigated to what degree each of the regimes imposed by the two laws has been successful in inducing a higher standard of care among firms in the hazardous waste industry.

In the remainder of this introduction, the economic issues facing industries in the hazardous substance and hazardous waste fields are reviewed. The economic costs imposed by RCRA regulations are discussed first, followed by a discussion of costs associated with liability and enforcement actions under RCRA and Superfund. The introduction concludes with a review of potential sources of future costs to industry.

II. REGULATORY COSTS

A. Generators and Transporters

Compliance with RCRA hazardous waste regulations has resulted in an increase in costs to most firms in one or more of the following areas:

1. Costs of Increased Administrative Burdens

Generators, transporters, storers, treaters, and disposers of hazardous wastes have to comply with complex sets of administrative requirements depending on the activities that they undertake. Generators and transporters incur costs of meeting operational requirements pertaining to record-keeping, labeling of hazardous waste containers, and conforming to the manifest system. Many generators also manage their wastes on-site and must therefore be fully aware of requirements for owners and operators of treatment, storage, and disposal facilities (TSOFs). Owners or operators of facilities that treat, store, or dispose of hazardous wastes incur additional administrative costs in meeting requirements for developing plans, procedures, and training programs; for reporting; and for securing facility permits. There may also be costs associated with reorganizing company management for compliance reasons.

2. Costs of Upgrading Facilities

Where an industry chooses to continue to manage (treat, store, or dispose of) its wastes on-site, costs will often be incurred to upgrade the facility to meet RCRA standards. These include both capital and operation and maintenance costs to meet the design and operational standards appropriate to the method of waste management practiced at the site.

For example, if a generator chooses to store wastes in drums prior to transfer to an off-site facility, he must appropriate drum storage facilities and certain handling procedures in place. Among these include the need for the container storage area to have a containment system designed to meet RCRA standards, and for proper inspections of the area to be performed weekly.

3. Cost of Avoiding Expensive Facility Upgrades

In many cases, a plant may choose to reduce wastes being generated to avoid the greater costs of upgrading the hazardous waste facility. Costs are also incurred in these waste reduction efforts. These include the costs of elimination of specific sources of waste, costs of product reformulation, costs of equipment redesign, and costs of marketing wastes for reuse and reclamation.

4. Costs of Off-Site Management

Many manufacturers do not have the capability of managing wastes on-site, nor do they want to develop this capability. These generators instead choose to send their wastes off-site for disposal. Costs have increased in many cases for these industries because they are sending more volumes of wastes off-site for disposal as a result of the RCRA regulations, and because the unit costs of off-site waste disposal (cost per drum of waste disposal) have also increased, as the commercial disposal industry responded to supply and demand pressures created by the regulations.

Virtually all generators and facility owners and operators have incurred some or all of these costs due to the RCRA hazardous waste regulations. These costs are likely to increase as EPA restricts the landfilling of wastes and imposes more stringent design and operating requirements in response to the most recent RCRA amendments. Some industries are better able to absorb these costs than others. Manufacturing facilities which operate at marginally profitable levels can be significantly affected by these added costs.

B. Treatment, Storage, and Disposal Facilities

RCRA requirements applicable to treatment, storage, and disposal facilities (TSDFs) can be classified under four broad categories:

1. Technical design standards
2. Financial responsibility requirements

3.   Permitting requirements
4.   Paperwork and administrative requirements

The economic effects of certain of these requirements will vary depending on the type of facility (e.g., storage vs. disposal), whether the facility was in operation prior to the promulgation of EPA's requirements, and the financial characteristics of the firm that owns the facility.

1. Technical Standards

Technical standards have been promulgated for all types of facilities, including container storage units, tank storage and treatment units, incinerators, and land-disposal facilities. In general, there are both design and operation standards that must be complied with.

As a result of the recent amendments to RCRA, land-disposal facilities in particular (defined to include landfills, surface impoundments, land treatment, and waste piles) are likely to incur large economic impacts. For example, land-disposal facilities unable to demonstrate compliance with ground-water monitoring and financial responsibility requirements were forced to cease operating on November 8, 1985. New landfills and surface improvements, including expansions to existing facilities, must be installed with two or more synthetic liners designed to prevent migration of hazardous constituents and leachate collection, leak detection, and ground-water monitoring systems. Moreover, existing surface impoundments must be retrofit to satisfy design standards for new facilities or be closed by 1988.

In addition to ground-water monitoring requirements applicable to all land-disposal facilities, all facilities must undertake extensive monitoring and perform corrective action measures if ground-water contamination is detected at the facility regardless of when the waste was disposed. Corrective action consists of the *removal* of the ground-water contamination (e.g., counterpumping) or *in situ* treatment of the hazardous constituents.

2. Financial Responsibility Requirements

EPA promulgated financial responsibility requirements for hazardous waste facilities in an attempt to prevent the abandonment or improper closure of hazardous waste facilities and their attendant hazards. All owners or operators of TSDFs that were in operation as of November 19, 1980 must ensure that funds will be available to cover the costs of closing the facility in an environmentally sound manner and maintaining it after closure, if it is a land disposal facility. Therefore, even if the owner or operator defaults on his obligations to close the facility properly, funds will be available so that closure can be completed by a third party. Owners or operators must also demonstrate financial responsibility for a limited level of third party liabilities. In response to Congressional mandate, EPA is now developing financial responsibility requirements to cover the costs of corrective action to remediate ground-water contamination.

The economic burden imposed by these financial responsibility requirements will vary significantly depending on the financial mechanism chosen to demonstrate financial responsibility. For firms that can satisfy the criteria of EPA's financial test, the costs of demonstrating financial responsibility for closure and post-closure and third party liability coverage are very low. In general, a significant number of firms with over $10 million in net worth are likely to pass EPA's financial test. For firms that cannot pass the test and must use one of the other allowable options (e.g., surety bond, letter of credit, trust fund, insurance), the costs of financial responsibility could be high. This is particularly true in light of the current constraints in the insurance market and the high costs of environmental impairment liability coverage. To help minimize the economic burden on firms that cannot pass a financial test, EPA allows a parent corporation that can pass a financial test to demonstrate financial responsibility for a subsidiary that would otherwise be required to use a higher priced financial instrument. EPA is now also considering allowing parent corporations the option of offering a corporate guarantee for liability requirements.

3. Permitting Requirements

The costs associated with obtaining a RCRA permit will vary depending on the type of operations being conducted at the facility, site-specific considerations (e.g., public opposition to siting) and whether the facility already has interim status or is applying for a permit to begin operations for the first time. Because land disposal facilities pose significantly higher risks than storage facilities, a prospective owner or operator of a land disposal facility will need to invest significant time and resources to obtain a permit (e.g., submittal of detailed engineering plans and supporting analyses and calculations, demonstration of financial viability, public hearings). Depending on the location of the proposed facility, significant resources may be needed to satisfy siting concerns.

The opportunity costs of obtaining a permit for a new facility may be unduly high if the firm owning the facility has no other source of revenues. If the facility is currently operating with interim status, this burden may not be so large. Although significant design changes may be required at the facility before a final permit is issued, in most cases an interim status facility will be allowed to continue operating and collecting revenues while a final permit is being negotiated.

4. Paperwork and Administrative Requirements

The EPA regulations contain a variety of reporting and record-keeping requirements including the following:

- Manifest system which tracks those hazardous wastes transported from the point of generation to an off-site disposal site
- Contingency plans
- Personnel training programs to ensure that employees know how to comply with the regulations especially in emergency situations
- Waste analysis plans for wastes being treated or disposed of at the facility
- Operating record of activities performed at TSDFs including type of wastes handled, manner of handling, location of disposal
- Air and ground-water monitoring, reporting, and record-keeping
- Closure and post-closure plans and cost estimates
- Financial responsibility reporting and record-keeping

In general, the paperwork requirements are not expected to pose a significant economic burden. Many of these administrative requirements correspond to good business practice (e.g., maintaining operating records, contingency plans, and personnel training programs). Other requirements involve up-front expenses with periodic low-cost updates subsequently (e.g., closure and post-closure plans and cost estimates, contingency plans).

III. LIABILITY COSTS

CERCLA provides an extensive liability scheme for persons responsible for the release or certain threatened releases of hazardous substances into the environment. Superfund imposes liability for response costs and natural resource damages on generators, transporters, and owners and operators of facilities from which there is a release or threatened release of hazardous substances. Covered facilities include, but are not limited to, hazardous waste disposal facilities. Thus, persons liable under CERCLA include not only the owner of a facility or vessel, but persons who arranged to have their hazardous substances disposed of or treated at the facility in question and transporters who selected the site in question for treatment or disposal.

Liability under CERCLA is limited to $50 million plus the costs of any necessary response. Defenses to liability include that the release was an act of God, an act of war, or an act of a third party with whom no contractual relationship existed. The statutory definitions of "facility", "release", "environment", and "hazardous substance" are broad. The liability scheme is broad and contains many incentives for private cleanup of hazardous substance releases, including the following:

- States have an independent federal cause of action against responsible parties.
- A responsible party under agreement with the government can clean up a multiple party site and use CERCLA's liability scheme to sue other responsible parties for reimbursement.
- Under certain circumstances, a third party can clean up a site and use CERCLA authority to sue responsible parties for reimbursement.
- The third party defense may be jeopardized if the otherwise responsible party fails to take mitigating actions in some circumstances.
- Limitations to liability are lost if the responsible party's actions constitute willful misconduct or willful negligence, or if the responsible party fails to take certain actions.
- Failure to take abatement action ordered by the Federal Government subjects the responsible party to possible treble punitive damages.
- Responsible parties are liable to the Federal and State Governments for injury to, destruction of, or loss of natural resources due to a release of hazardous substances.

RCRA does not establish a commensurate liability scheme; it provides a narrower range of remedies, applies generally to fewer substances, and is available in a narrower range of situations. Nonetheless, sanctions that can be brought under RCRA are not insignificant incentives for responsible parties to address hazardous waste problems. When Subtitle C provisions are violated, there are a variety of sanctions that can be imposed, depending on the circumstances, that include: (1) revocation of permit; (2) injunctive action by the courts; (3) fines of $25,000 to $50,000/day, plus flat fines of $250,000 to one million dollars where there is knowing endangerment of life or threat of bodily harm; and (4) up to 2 years imprisonment.

Both CERCLA and RCRA authorize enforcement actions when an "imminent and substantial endangerment" exists. Under Section 7003 of RCRA, the EPA Administrator may (1) bring suit in U.S. District Court to obtain a restraining order against the persons contributing to the endangerment or (2) issue administrative orders or take other actions necessary to protect public health and the environment. The provisions of RCRA Section 7003 are triggered

> upon receipt of evidence that the handling, storage, treatment, transportation, or disposal of any solid waste or hazardous waste may present an inmminent and substantial endangerment to health or the environment.

Failure to comply with the Administrator's orders can result in a $5000/day fine.

CERCLA Section 106 is similar to RCRA Section 7003. CERCLA authorizes suit by the Attorney General to secure the relief required by the public interest and the equities of a particular case. The President is authorized (which authority has been delegated to the EPA Administrator) to issue orders to protect the public health or welfare or the environment. Action under CERCLA Section 106 is triggered when a determination is made that "there may be an imminent and substantial endangerment to the public health or welfare or the environment." Failure to comply with the Administrator's orders subjects the owner/operator, disposer, or transporter to a fine of $5,000/ day of noncompliance and may subject that person to treble punitive damages in an action by the government to recover its response costs (see CERCLA Section 107(c)(3)).

CERCLA's imminent and substantial endangerment provision applies retroactively. That is, an owner/operator, transporter, or disposer may be subject to an order to abate conditions resulting from actions taken by that person before CERCLA was enacted. Thus, although such a person may have complied with the relevant law at the time the action in question was taken, that person still must comply with an administrative or court order issued under CERCLA Section 106. Whether RCRA's imminent and substantial endangerment provision is also retroactive is less clear; there are conflicting federal court decisions on the matter. One district court has specifically held that RCRA Section 7003 applies only prospectively, but other courts have held that the section applies retroactively as well. Retroactive application of RCRA Section 7003 would significantly expand the number of persons and sites subject to its abatement provisions.

IV. FUTURE ECONOMIC ISSUES

Future costs on industry involved in hazardous substances and hazardous waste may arise as a result of changes in the RCRA regulations, implementation of the Post-Closure Liability Trust Fund, and the possibility of legislation in the area of victim compensation.

As the RCRA program evolves, attention will inevitably shift from regulation of the transportation, handling, and disposal of hazardous wastes at active facilities to the post-closure care of disposal facilities. Additional sources of costs can be expected to arise as greater understanding develops of the requirements for proper closure and post-closure care.

Another focus of economic issues in the future is the Post-Closure Liability Trust Fund established by Section 232 of CERCLA. Imposition of the Post-Closure Fund tax began in September 1983. The Fund is financed by a tax of $2.13 per dry weight ton of hazardous waste received at a qualifying hazardous waste facility. The tax shuts off when the unobligated balance in the Fund in the preceding year exceeds $200 million.

The Post-Closure Fund assumes the liability of the owner and operator of a hazardous waste disposal facility under all federal and state law, provided that certain conditions have been met. The facility must have received a RCRA general status permit; the facility must have been closed in compliance with the conditions of its permit and applicable regulations relating to closure or affecting the performance of the facility after closure; and the facility and the surrounding area must be monitored for up to 5 years following closure to demonstrate that there is no substantial likelihood of a release of hazardous substances or of other risk to public health or welfare.

Given these conditions, it is unlikely that transfers of liability to the Post-Closure Fund will begin for several more years. As of early 1986, the status of the Post-Closure Fund remains unclear. No implementing regulations have been promulgated, the Fund has not been invoked for a single site, and the various Superfund reauthorization bills either abolish the Post-Closure Fund or require additional study before the future of the Fund is determined.

Although it is not possible to determine what the ultimate configuration of the Post-Closure Fund might be, the statutory provisions and any regulations that EPA develops could have a major effect on who assumes responsibility for the continual care of inactive hazardous waste sites in perpetuity.

Finally, another related potential source of future costs to industry lies in the possibility that some form of victim compensation legislation will be enacted by Congress. The subject of victim compensation has received considerable discussion, and the full extent of its effects on industry would clearly depend on the particular type of legislation under consideration. Nevertheless, most forms of victim compensation legislation could have a major effect on industry actions under RCRA and Superfund. As a simple example, if victim compensation provisions are enacted as amendments to CERCLA, there would be a greatly increased urgency for a firm to obtain a RCRA facility permit in order to forestall liability for any releases, since federally permitted releases are excluded from the liability imposed by CERCLA.

V. SUMMARY

Over the last several years, a RCRA/Superfund framework of liability, enforcement, permits, and regulations has been put in place with the goal of fundamentally changing the management of hazardous substances and hazardous wastes. The full costs of implementing that framework are still being incurred, particularly as the RCRA permit program gets underway. Attention can be expected to shift over the next few years to the post-closure care of hazardous waste facilities and to the issues involved in compensating victims of hazardous substance releases.

TOXIC TORT REMEDIES

1. Baurer, T., Love Canal: Common Law Approaches to a Modern Tragedy, 11 *Envt'l L.* 133 (1981).

1a. Bayer, D., Joint and Several Liability: Easing the Plaintiff's Burden in Toxic-Tort Cases, 21 Trial Feb. 1985 at 56-64.

2. Belfiglio, J., Hazardous Waste: Preserving the Nuisance Remedy, 33 *Stan. L. Rev.* 675 (1981).

3. Bruno, M., The Development of a Strict Liability Cause of Action for Personal Injuries Resulting from Hazardous Waste, 16 *New Engl. L. Rev.* 543 (1980-81).

3a. Buckley, C., A Suggested Remedy for Toxic Injury: Class Actions, Epidemiology and Economic Efficiency 26 Wm. & Mary L. Rev. 497-543 (Spring 1985).

4. Fabic, M., Pursuing a Cause of Action in Hazardous Waste Pollution Cases, 29 *Buffalo L. Rev.* 533 (1980).

5. Feinberg, D., Denial of a Remedy: Former Residents of Hazardous Waste Sites and New York's Statute of Limitations, 8 *Colum. J. Envt'l L.* 161 (1982).

6. French, B., Private Nuisance Approach to Hazardous Waste Disposal Sites, 7 *Ohio N.U.L. Rev.* 86 (1980).

7. Fulton, C., Hazardous Waste: Third Party Compensation for Contingencies Arising from Inactive and Abandoned Hazardous Waste Disposal Sites, 33 *S.C.L. Rev.* 543 (1982).

8. Garrett, T., Compensating Victims of Toxic Substances: Issues Concerning Proposed Federal Legislation, 13 *Envt'l L. Rep.* 10172 (1983).

9. Ginsberg, W. and Weiss, L., Common Law Liability for Toxic Torts: A Phantom Remedy, 9 *Hofstra L. J.* 859 (1983).

10. Hinds, R., Liability under Federal Law for Hazardous Waste Injuries, 6 *Harv. Envt'l L. Rev.* 1 (1982).

11. Hornabach, D., Toxic Torts — Is Strict Liability Really the Fair and Just Way to Compensate the Victims?, 16 *U. Rich. L. Rev.* 305 (1982).

12. Hurwitz, W., Environmental Health: An Analysis of Available and Proposed Remedies for Victims of Toxic Waste Contamination, 7 *Am. J. L. & Med.* 61 (1981).

13. Jernberg, D., Insurance for Environmental and Toxic Risks: A Basic Analysis of the Gap Between Liability and Coverage, 34 *Fed'n Ins. Couns.* 103 (1984).

14. Kelly, M., What to do When the Deep Pocket Goes Under, 69 *A.B. A.J.* 740 (1983).

15. Light, A., Toxic Injury Compensation: One More Overview of Legislative Proposals, 15 *Nat. Resources L. Newsl.* 2(5) (1983).

15a. Maynard, R. And Crisci, G., the Duty to Warn in "Toxic Tort" Litigation 33 Clev. St. La. Rev. 69 (1984-85).

16. McGovern, F., Toxic Substances Litigation in the Fourth Circuit, 16 *U. Rich L. Rev.* 247 (1982).

17. Meyer, S., Compensating Hazardous Waste Victims: RCRA Insurance and a Not So "Super" Fund Act, 11 *Envt'l L.* 689 (1981).

18. Miller, J., Private Enforcement of Federal Pollution Control Laws, Part I, 13 *Envt'l L. Rep.* 10309 (1983); Part II, 14 *Envt'l L. Rep.* 10063 (1984).

19. Miller, R., Initial Processing in the Toxic Tort Case, 20 *Trial* 71-3 (1984).

20. Mulcahy, M., Proving Causation in Toxic Torts Litigation, 11 *Hofstra L. Rev.* 1299 (1983).

21. Note, The Fairness and Constitutionality of Statutes of Limitations for Toxic Tort Suits, 96 *Harv. L. Rev.* 1683 (1983).

22. Obremski, C., Toxic Tort Litigation and the Insurance Coverage Controversy, 34 *Fed'n Ins. Counsel Q.* 3 (1983).

23. Parnell, A., Manufacturers of Toxic Substances: Tort Liability and Punitive Damages, 17 *Forum* 947 (1982).

24. Schmidt, S. et al., The Constitutionality of Federal Products Liability — Toxic Tort Legislation, 6 *J. Products Liability* 171 (1983).

25. Singer, S., An Analysis of Common Law and Statutory Remedies for Hazardous Waste Injuries, 12 *Rutgers L. J.* 117 (1980).

26. Soble, J., A Proposal for the Administrative Compensation of Victims of Toxic Substance Pollution: A Model Act, 14 *Harv. J. Legislation* 683 (June, 1977).

27. Sokolow, M., Hazardous Waste Liability and Compensation: Old Solutions, New Solutions, No Solutions, 14 *Conn. L. Rev.* 307 (1982).

28. Stanley, V., Establishing Liability for the Damages from Hazardous Wastes: An Alternative Route for Love Canal Plaintiffs, 31 *Cath. U. L. Rev.* 273 (1982).

29. Strand, P., The Inapplicability of Traditional Tort Analysis to Environmental Risks: The Example of Toxic Waste Pollution Victim Compensation, 35 *Stan L. Rev.* 575 (1983).

30. Trauberman, J. et al., Compensation for Toxic Substances Pollution, Michigan Case Study, 10 *Envt'l L. Rep* 50021 (1980).

31. Trauberman, J., Compensating Victims of Toxic Substances Pollution: An Analysis of Existing Federal Statutes, 5 *Harv. Envt'l L. Rev.* (1981).

32. Trauberman, J., Statutory Reform of Toxic Torts: Relieving Legal, Scientific and Economic Burdens on the Chemical Victim, 7 *Harv. Envt'l Rev.* 177 (1983).

33. Tredway, L., When a Veteran Wants Uncle Sam: Theories of Recovery for Servicemembers Exposed to Hazardous Substances, 31 *Am. U. L. Rev.* 1095 (1982).

34. Zinns, J., Close Encounters of the Toxic Kind — Towards an Amelioration of Substantive and Procedural Barriers for Latent Toxic Injury Plaintiffs, 54 *Temp. L. Q.* 822 (1981).

RCRA

1.   Anderson, R., The Resource Conservation and Recovery Act of 1976: Closing the Gap, 1978 *Wis. Law Rev.* 633.

2.   Bartlett, K., The Constitutional Framework of RCRA, *Toxic Substances J.,* Vol. 3 (1981).

3.   Buc, L. et al., Regulating Hazardous Waste Incinerators Under the Resource Conservation and Recovery Act, 23 *Nat. Resources J.* 549 (1983).

4.   Cohen, N., Landowner Liability Under Common Law, RCRA, The Clean Water Act, The Refuse Act and CERCLA, 2 *Chem. & Rad. Waste Litg. Rep.* 402 (1981).

5.   Costle, D., Environmental Regulation and Regulatory Reform, 57 *Wash. L. Rev.* 409 (1982).

6.   Deutch, A. et al., An Analysis of Regulations Under the Resource Conservation and Recovery Act, 25 *Wash. U. J. Urb, & Contempt L.* 145 (1983).

6a. DiBenedetto, J., Generator Liability under the Common Law and Federal and State Statutes 39 Bus. Law 611-638 (February 1984).

7.   Duke, K., Using RCRA: Iminent Hazard Provision in Hazardous Waste Emergencies, 9 *Ecology L. Q.* 599 (1981).

8.   Flannery, D. and Poland, K., Hazardous Waste Management Act: Closing The Circle, 85 *W. Va. L. Rev.* 347 (1982).

9.   Friedland, S., The New Hazardous Waste Management System: Regulation of Wastes or Wasted Regulation?, 5 *Harv. Envt'l L. Rev.* 89 (1981).

9a. Gutierrez and Bullerdick, RCRA's Underground Storage Tanks and Corrective Action Provisions, Envt'l Forum, September 1985 at 33.

9b. Harrington, A., The 1984 Amendments to the Resource Conservation and Recovery Act, 58 Wis. B. Bull. 17 (4) June 1985.

10. Kovacs, W. and Klucsik, J., The New Federal Role in Solid Waste Management: The Resource Conservation and Recovery Act of 1976, 3 *Colum. J. Envt'l L.* 205 (1977).

11. Laswell, D., State - Federal Relations Under Subtitle C of the Resource Conservation and Recovery Act, 16 *Nat. Resources Law* 641 (1984).

12. Lee, J., RCRA's State Program Provisions and the Tenth Amendment: Coercion or Cooperation, 9 *Ecology L. Q.* 579 (1981).

13. Marten, B., Regulation of the Transportation of Hazardous Materials: A Critique and a Proposal, 5 Harv. *Env'tl L. Rev.* 245 (1981).

14. Marnell, M., EPA's Responsibilities Under RCRA: Administrative Law Issues (Hazardous Substances in the Environment), 9 *Ecology L.Q.* 555 (1981).

14a. Rasbe, W. and Galley, R., The Hazardous and Solid Waste Amendments of 1984: A Dramatic Overhaul of the Why America Manages Its Hazardous Wastes, 14 Envt'l L. Rev. 10458-10467 (Dec. 1984).

15. Rosbe, W., RCRA and Regulation of Hazardous and Nonhazardous Solid Wastes — Closing the Circle of Environmental Control, 35 *Bus. Law.* 1519 (1980).

16. Ruda, J. and Hoffman, M., Overview of Hazardous Waste Regulation Under RCRA, 10 *Colo. Law* 234 (1981).

17. Schnapf, D., State Hazardous Waste Programs Under the Federal Resource Conservation and Recovery Act, 12 *Envt'l L.* 679 (1982).

18. Skilling, K., Solid Wastes Programs and the Resource Conservation and Recovery Act of 1976, *BNA Environ. Rep.* Vol. 8, No. 22 (monograph No. 6) (October 7, 1977).

19. Trilling, B., Potential for Harm as the Enforcement Standard for Section 7003 of the Resource Conservation and Recovery Act (Symposium: Hazardous Substances), 2 *UCLA J. Env'tl L. & Pol'y* 43 (1981).

20. Tripp, J. and Jaffe, A., Preventing Groundwater Pollution: Towards a Coordinated Strategy to Protect Critical Recharge Zones, 3 *Harv. Env'tl L. Rev.* 1 (1979).

20a. Want, W., Understanding the New RCRA . . . Minimum Technological Requirements, Envt'l. Forum, July 1985 at 25.

21. Weiland, R., Enforcement Under the Resource Conservation and Recovery Act of 1976, 8 *B.C. Env'tl Aff. L. Rev.* 641 (1980).

22.	Worobec, R., Analysis of the Resource Conservation and Recovery Act, *BNA Environ. Rep.* (Current Developments), Vol. 11, No. 17 (August 22, 1980).

FEDERALISM

1.	Anson, T. and Schenkkan, P., Federalism, the Dormant Commerce Clause, and State-owned Resources, 59 *Tex. L. Rev.* 71 (1980).
2.	Babbitt, B., Federalism and the Environment: A Change in Direction, 12 *Envt'l L.* 847 (1982).
3.	Banks, W., Conservation, Federalism and the Courts: Limiting the Judicial Role, 34 *Syracuse L. Rev.* 685 (1983).
4.	Bleiweiss, S., Environmental Regulation and the Federal Common Law of Nuisance: A Proposed Standard of Preemption, 7 *Harv. Envt'l L. Rev.* 41 (1983).
5.	Cain, J., Routes and Roadblocks: State Controls on Hazardous Waste Imports, 23 *Nat. Resources J.* 767 (1983).
6.	Campbell, W., State Ownership of Waste Disposal Sites: A Technique for Excluding Out-of-State Wastes?, 14 *Envt'l L.* 177 (1983).
7.	Comment, State Environmental Protection Legislation and the Commerce Clause, 87 *Harv. L. Rev.* 1762 (1974).
8.	Danaher, J., Waste Embargo Held a Violation of the Commerce Clause, 11 *Conn. L. Rev.* 292 (1979).
9.	Fairfax, S., Old Recipes for New Federalism (Symposium: Federalism and the Environment: A Change in Direction), 12 *Envt'l L.* 948 (1982).
10.	Florini, K., Issues of Federalism in Hazardous Waste Control: Cooperation or Confusion?, 6 *Harv. Envt'l L. Rev.* 307 (1982).
11.	Flynn, P., The Constitutionality of State Environmental Laws Under the Commerce Clause: City of Philadelphia v. New Jersey, 5 *Envir. Affairs* 721 (1976).
11a.	Laswell, D., State — Federal Relations under Subtitle the Resource Conservation and Recovery Act. 16, Nat. Resources Law. 641-664 (Winter 1984).
12.	Leman, C. and Nelson, R., The Rise of Managerial Federalism: An Assessment of Benefits and Costs, 12 *Envt'l L.* 981 (1982).
13.	Lester, A., Decisions and Actions of 1983 Affecting Federalism, 54 *Okla. B.J.* 3282 (1983).
14.	Lutz, R., Interstate Environmental Law: Federalism Bordering on Neglect?, 13 *Sw. U.L. Rev.* 571 (1983).
15.	Lyons, W., Federalism and Resource Development: A New Role for States?, 12 *Envt'l L.* 931 (1982).
16.	Renz, J., The Effect of Federal Legislation on Historical State Powers of Pollution Control: Has Congress Muddied State Waters?, 43 *Mont. L. Rev.* 197 (1982).
17.	Rhodes, R. and MacLaughlin, D., Federal and State Regulation of Hazardous Waste Management (Florida), 54 *Fla. B.J.* 713 (1980).
18.	Stever, D., Deference to Administrative Agencies in Federal Environmental Health and Safety Litigation — Thoughts on Varying Judicial Application of the Rule, 6 *W. New Engl. L. Rev.* 35 (1983).
18a.	Warren, E., State Hazardous Waste Superfunds and CERCLA: Conflict or Complement? 13 Envt'l L. Rep. 10348 — 10360 (Nov. 83).

HAZARDOUS WASTE

1.	Brenner, J., Liability for Generators of Hazardous Waste: The Failure of Existing Enforcement Mechanisms, 69 *Geo. L.J.* 1047 (1981).
2.	Eckhardt, R., The Unfinished Business of Hazardous Waste Control, 33 *Baylor L. Rev.* 253 (1981).
3.	Gentges, A., Hazardous Waste Injection Wells: The Need for State Controls, 19 *Tulsa L. J.* 250 (1983).
4.	Hall, K., Health Risks from Exposure to Hazardous Wastes, 14 *Envt'l L. Rep.* 10118 (1984).
5.	Hall, R., The Problem of Unending Liability for Hazardous Waste Management, 38 *Bus. Law* 593 (1983).
6.	Hedeman, W., Public Concern of Hazardous Waste Sites, 14 *Envt'l L. Rep.* 10116 (1984).
7.	Joest, G., Will Insurance Companies Clean The Augean Stables? Insurance Coverage for the Landfill Operator, 50 *Ins. Counsel J.* 258 (1983).
8.	Jorling, T., Hazardous Substances in the Environment, 9 *Ecology L.Q.* 520 (1981).
9.	Note, Allocating the Costs of Hazardous Waste Disposal, 94 *Harv. L. Rev.* 584 (1981).
10.	Rodburg, M., Land Ownership and Hazardous Waste Law, *N.J. Law* 12 (1983).
11.	Rudolph, S., Case Comment of *Village of Wilsonville* v. *SCA Services*, 70 *Ill. B.J.* 586 (1982).
12.	Sussna, S., Remedying Hazardous Waste Facility Siting Maladies by Considering Zoning and other Devices, 16 *Urb. L.* 29 (1984).

13. Tarlock, A., State v. Local Control of Hazardous Waste Facility Siting: Who Decides in Whose Backyard, 7 *Zoning and Plan L. Rep.* 9 (1984).

14. Trilling, B., Painstaking Negotiation Leads to Landmark Court Order Approving Settlement Agreement in Hyde Park Hazardous Waste Cleanup Litigation, 12 *Envt'l L. Rep.* 15103 (1982).

15. Wilhelm, G., The Regulation of Hazardous Waste Disposal: Cleaning the Augean Stables with a Flood of Regulations, 33 *Rutgers L. Rev.* 906 (1981).

SUPERFUND

1. Anderson, C., Superfund Proposed to Clean up Hazardous Waste Disasters, 20 *Nat. Resources J.* 615 (1980).

2. Anderson, C., Hazardous Wastes: Superfund Solution?, 1 *Wm. Mitchell Envt'l L.J.* 162 (1983).

2a. Anderson, F., Negotiation and Informal Agency Action: The Case of Superfund, Dulce L. Jour. April 1985 pp. 261-380.

2b. Anderson, K., The Plight to Contribution for Response Costs Under CERCLA, 60 Notre Dame L. Rev. 345 (1985).

2c. Baskin, S. and Reed, P., "Arranging for Disposal" Under CERCLA: When is a Generator Liable? 15 ELR 10160 (1985).

3. Benik, G., The Environmental Crisis of Hazardous Waste — Superfund: A Congressional Response, 17 *Ark. Law* 100 (1983).

4. Bernstein, N., The Enviro-Chem Settlement: Superfund Problem Solving, 13 *Envt'l L. Rep.* 10402 (1983).

5. Brew, B., Natural Resource Recovery by Federal Agencies — A Road Map to Avoid Losing Causes of Action, 13 *Envt'l L. Rep.* 10324 (1983).

6. Chesler, A., Clean-up of Hazardous Waste Sites Under Superfund — Considerations of Corporate Counsel, 18 *Law Notes* 115 (1982).

7. Comment, State Hazardous Waste Superfunds and CERCLA: Conflict or Complement?, 13 *Envt'l L. Rep.* 10348 (1983).

8. Dellecker, R., The Pre-emptive Scope of the Comprehensive Response, Compensation and Liability Act of 1980: Necessity for an Active State Role, 34 *U. Fla. L. Rev.* 635 (1982).

9. Dore, M., The Standard of Civil Liability for Hazardous Waste Disposal Activity: Some Quirks of Superfund, 57 *Notre Dame Law* 260 (1981).

10. Freeman, G., Toxic Torts, Hazardous Waste and the Superfund, 2 *J. Prod. L.* 149 (1983).

11. Grad, F., Hazardous Waste Victim Compensation: The Report of the Section 301(e) Superfund Study Group; A Response to Theodore L. Garnett, 13 *Envt'l L. Rep.* 10235 (1983).

12. Grad, F., Injuries from Exposure to Hazardous Waste: Can the Victim Recover? A Comment on the Report of the Section 301(e) Superfund Study Group, 2 *J. Prod. L.* 133 (1983).

13. Grad, F., A Legislative History of the Comprehensive Environmental Response Compensation and Liability *"Superfund")* Act of 1980, 8 *Colum. J. Envt'l L.* 1 (1982).

14. Grad, F., Remedies for Injuries Caused by Hazardous Waste: The Report and Recommendations of the Superfund 301(e) Study Group, 14 *Envt'l L. Rep.* 10105 (1984).

15. Gulick, T., Superfund: Conscripting Industry Support for Environmental Cleanup (Hazardous Substances in the Environment), 9 *Ecology L. Q.* 524 (1981).

16. Light, A., A Comparison of the 301(e) Report and Some Pending Legislative Proposals, 14 *Envt'l L. Rep.* 10133 (1984).

16a. Light, A., A Defense Counsel's Respective on Superfund, 15 ELR 10203 (July 1985).

17. Luster, E., The Comprehensive Environmental Response, Compensation and Liability Act of 1980: Is Joint and Several Liability the Answer to Superfund?, 18 *New Eng. L. Rev.* 109 (1983).

17a. Luster, E., The Comprehensive Environmental Response, Compensation and Liability Act of 1980: Is Joint and Several Liability the answer to Superfund? 18 New Eng. L. Rev. 109-147 L, Winter 1983).

18. Macbeth, A. and Mayer, R., An Introduction to Superfund, 30 *Prac. Law* 53 (1984).

18a. Macbeth, A. and Mayer, R., An Introduction to Superfund 30. *Prac. Law* 53-72 March 1, 1984.

19. Marzulla, R., The Government Response to the Environmental Defense Fund/Chemical Manufacturers Association Section 104(i) Litigation, 14 *Envt'l L. Rep.* 10120 (1984).

20. Menefee, M., Recovery for Natural Resource Damages Under Superfund: The Role of the Rebuttal Presumption, 12 *Envt'l L. Rep.* 15057 (1982).

21. Mintz, J., A Response to Rogers, Three Years of Superfund, 14 *Envt'l L. Rep.* 10036 (1984).

22. Mott, R., Defenses Under Superfund, 13 *Nat. Resources L. Newsl.* 1 (1981).

23. Mott, R., "How Dirty is Clean?", A Comparison of Settlement Devices for Defining the Appropriate Extent of Remedial Action by the Defendant, 5 *Chem. & Rad. W. Litg. Rep.* 1176 (1983).

24. Nathan, I. and Weiner, R., Superfund for Asbestos Liabilities: A Sensible Solution to a National Tragedy, 14 *Envt'l L. Rep.* 10127 (1984).

25. Note, Joint and Several Liability for Hazardous Waste Releases Under Superfund, 68 *(Va. L. Rev.* 1157 (1982).

25a. Note, Private Cost Recovery Actions under CERCLA, 69 *Minn. L. Rev.* 1135 (1985).

26. Pain, G., Mega-Party Superfund Negotiations, 12 *Envt'l L. Rep.* 15054 (1982).

26a. Parker, W., Chemical Industry Pushes for Broader Based Superfund Tax. 26 Tax Note 1191 March 25, 1985.

27. Pollock, A., The Role of Injunctive Relief and Settlements in Superfund Enforcement, 68 *Cornell L. Rev.* 706 (1983).

27a. Reed, P., CERCLA litigation update: The Emerging Law on Generator Liability. 14 *Envt'l Law Rep.* 10224-10236 June (1984).

28. Rikleen, L., Negotiating Superfund Settlement Agreements, 10 *B.C. Envt'l Aff. L. Rev.* 697 (1983).

28a. Rikleen, L., Superfund Settlements: Key to Accelerated Waste Cleanups, *Enuth. Forum August* 1985 at 51.

29. Rogers, J., The Generator's Dilemma in Superfund Cases, 12 *Envt'l L. Rep.* 15049 (1982).

30. Rogers, J., Three Years of Superfund, 13 *Envt'l L. Rep.* 10361 (1983).

31. Stoll, R., Litigation to Force CERCLA Health Studies, 15 *Nat. Resources L. Newsl.* 3(2) (1983).

32. Stoll, R., The 104(i) Litigation and the Chemical Industry's Concerns about Recent Compensation Proposals, 14 *Envt'l L. Rep.* 10119 (1984).

33. Thomas, F., Municipal and Private Party Claims Under Superfund, 13 *Envt'l L. Rep.* 10272 (1983).

34. Thomas, F., Superfund and the Eleventh Amendment: Are the States Immune from Section 107 Suits?, 14 *Envt'l L. Rep.* 10156 (1984).

34a. Want, W., CERCLA Amendments — The House Subcommittee Bill, 15 ELR 10200 (1985).

35. Zazzali, J. and Grad, F., Hazardous Wastes: New Right and Remedies? The Report and Recommendations of the Superfund Study Group, 13 *Seton Hall L. Rev.* 446 (1983).

INDEX